# CRICKET IN THE GRASS

# CRICKET IN THE GRASS

# AND OTHER STORIES

by Philip Van Soelen

Sierra Club Books / Charles Scribner's Sons
SAN FRANCISCO / NEW YORK

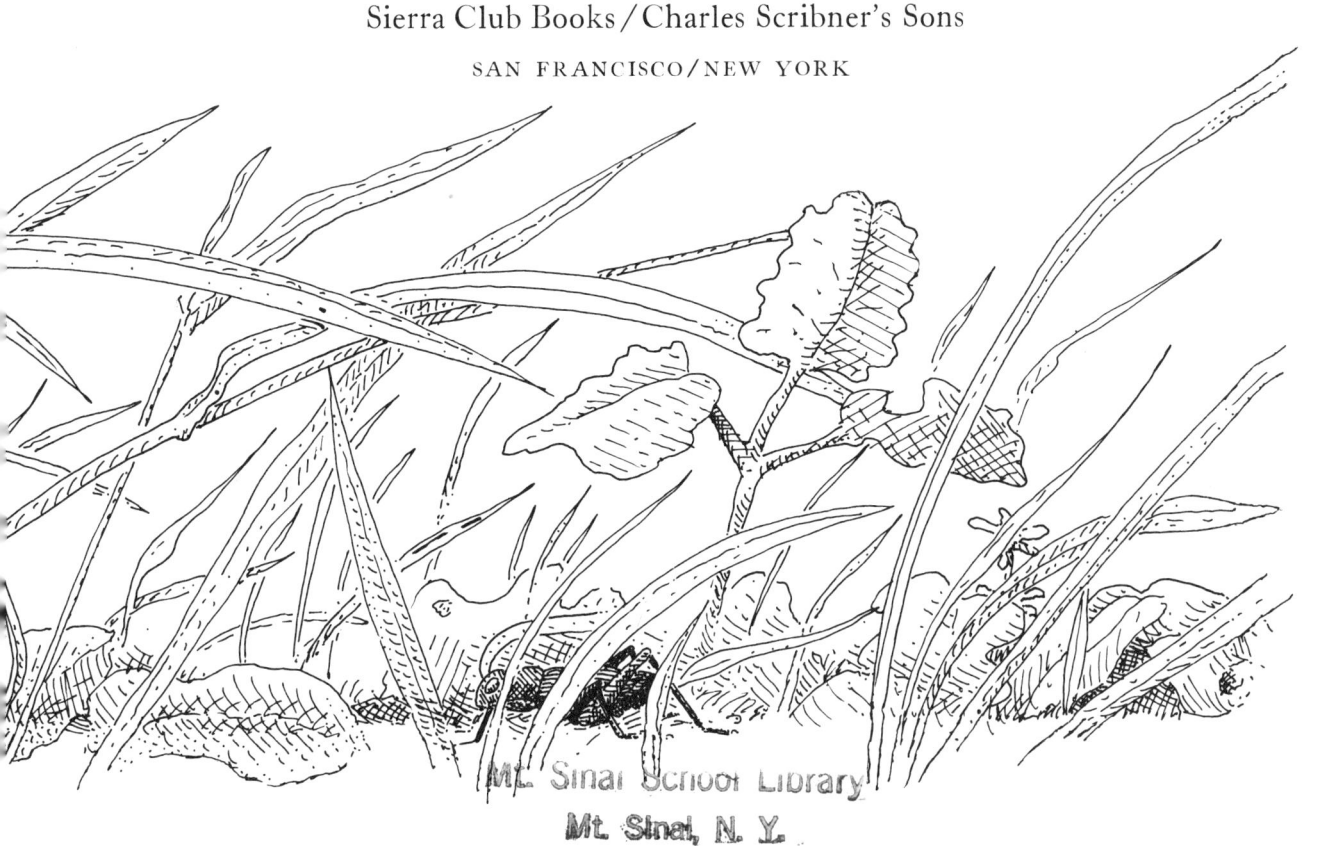

A YOLLA BOLLY PRESS BOOK

Copyright © 1979 by Philip Van Soelen

All rights reserved. No part of this book may be reproduced in any form or by any electronic or mechanical means including information storage and retrieval systems without permission in writing from the publisher. Trade distribution is by Charles Scribner's Sons, 597 Fifth Avenue, New York, New York 10017.

*Cricket in the Grass* was edited and prepared for publication at The Yolla Bolly Press, Covelo, California, during the winter of 1978-79 under the supervision of James and Carolyn Robertson. Production staff: Joyca Cunnan, Carla Shafer, Diana Fairbanks, Barbara Speegle, and Dan Hibshman.

The Sierra Club, founded in 1892 by John Muir, has devoted itself to the study and protection of the earth's scenic and ecological resources — mountains, wetlands, woodlands, wild shores and rivers, deserts and plains. Its publications are part of the nonprofit effort the club carries on as a public trust. There are some 50 chapters coast-to-coast, in Canada, Hawaii, and Alaska. For information about how you may participate in the club's program to enjoy and preserve wilderness and the quality of life, please address inquiries to Sierra Club, 530 Bush Street, San Francisco, California 94108.

Manufactured in the United States of America

1 3 5 7 8 11 13 15 17 19 MD/C 20 18 16 14 12 10 8 6 4 2
1 3 5 7 9 11 13 15 17 19 MD/P 20 18 16 14 12 10 8 6 4 2

Library of Congress Cataloging in Publication Data

Van Soelen, Philip, 1951-
Cricket in the grass.
SUMMARY: Five interconnected stories told primarily through illustrations reveal the teeming life and often sudden death occuring in a watershed.
1. Ecology — Juvenile literature.
[1. Ecology — Pictorial works]   I. Title.
QH541.14.V36        574.5        79-4108
ISBN 0-684-16110-9

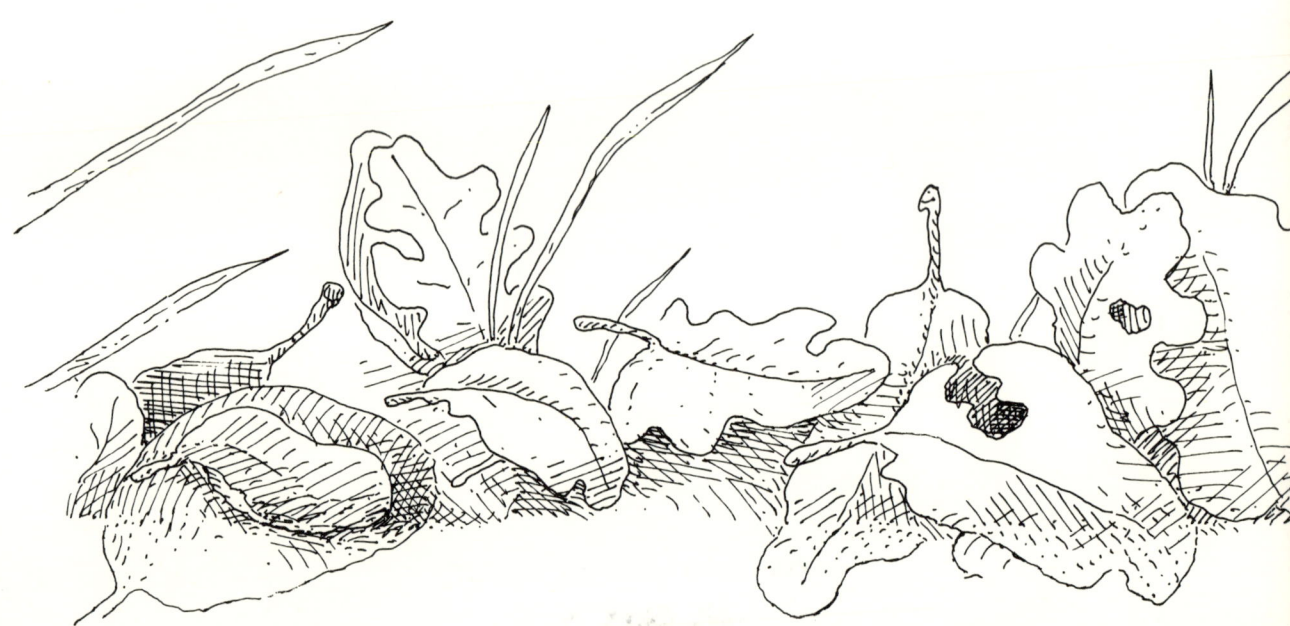

FOR ANNIE

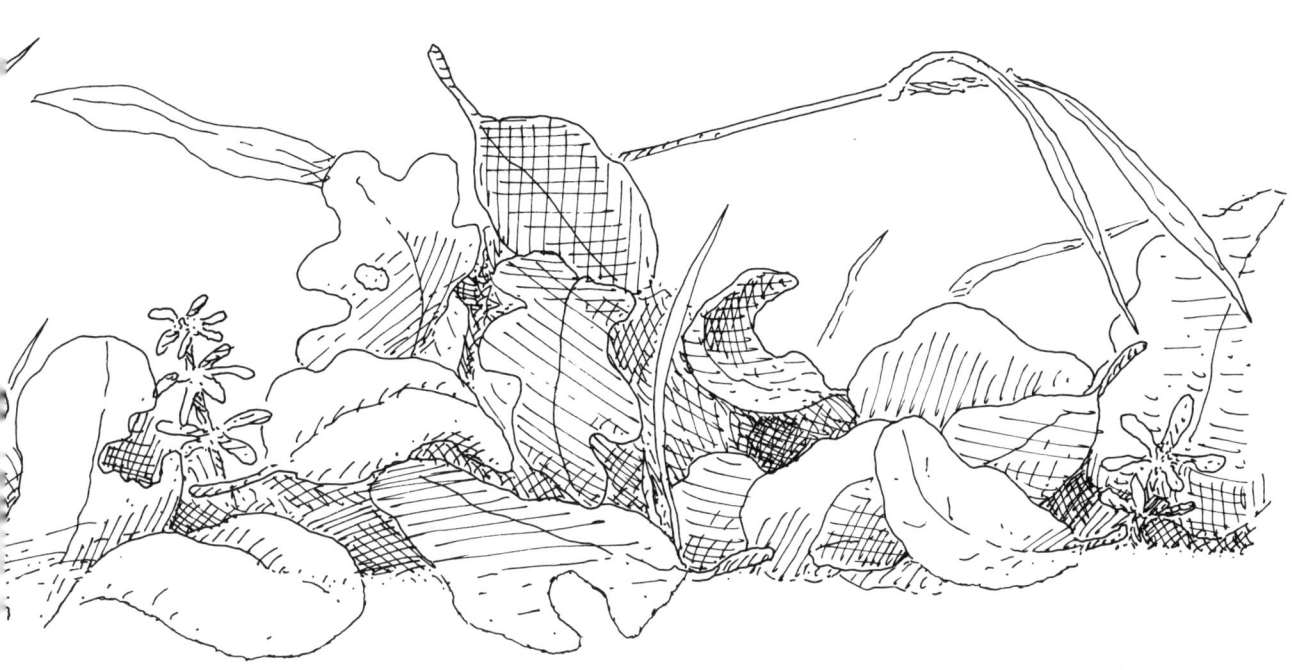

*This is a book about living places.* Places where animals and plants live are called habitats. There are many habitats in the world. There are dark wet habitats, dry warm habitats, and light grassy ones. Almost every place on the earth is a home for some creature. Even the deepest oceans and the highest mountains are homes for living things. If a plant or animal is well suited to its living place, it is said to be well adapted. Creatures and plants that are well adapted to their habitats are apt to survive in them for a long time.

The stories in this book are told with drawings. If you look carefully, you will see that each story has many parts. After looking, you may have some questions. At the back of the book are some words, and there you may find answers.

# THE STORIES

I   CRICKET IN THE GRASS, page 11
Who Lives in the Grassland?

II   THE OAK LOG, page 35
Soil Makers and Dwellers

III   THE GREEN LEAF CREATURES, page 49
At Home on a Sow Thistle

IV   A TURTLE'S WORLD, page 63
Living in the Stream

V   THE SALT MARSH, page 79
Meeting of the Waters

ABOUT THE STORIES, page 103
Some Words About the Pictures

# I
# CRICKET IN THE GRASS

# II
# THE OAK LOG

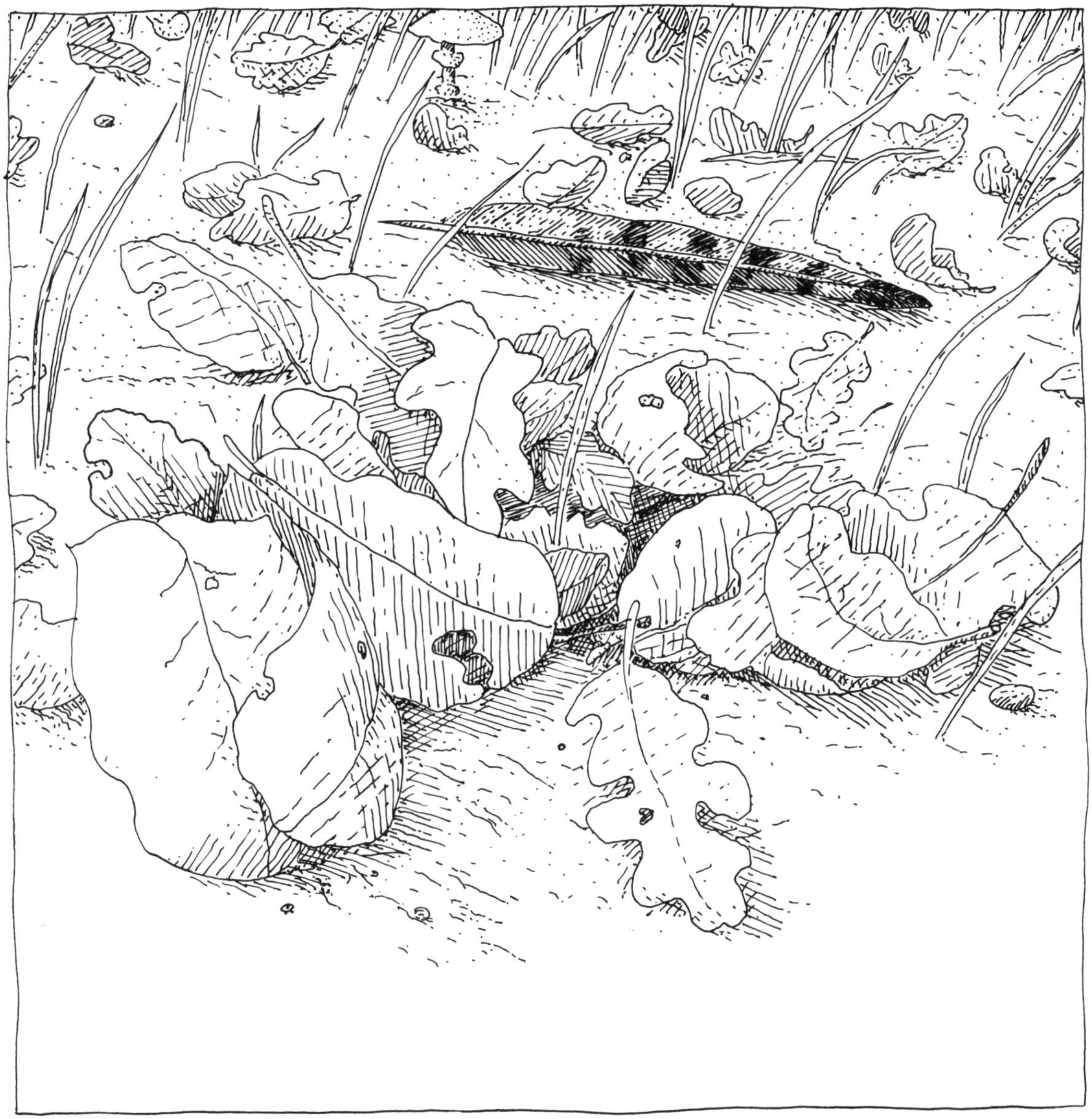

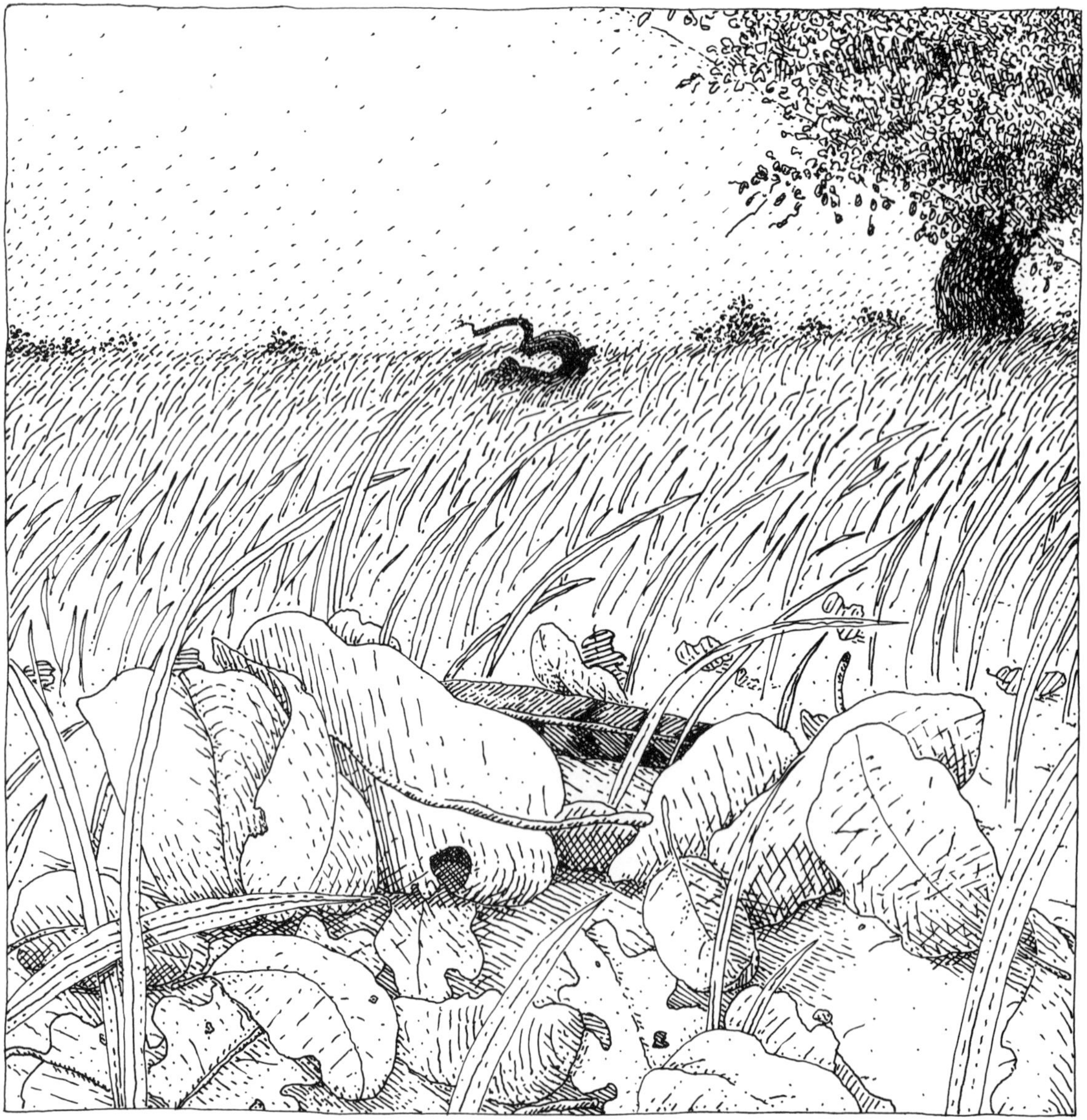

39

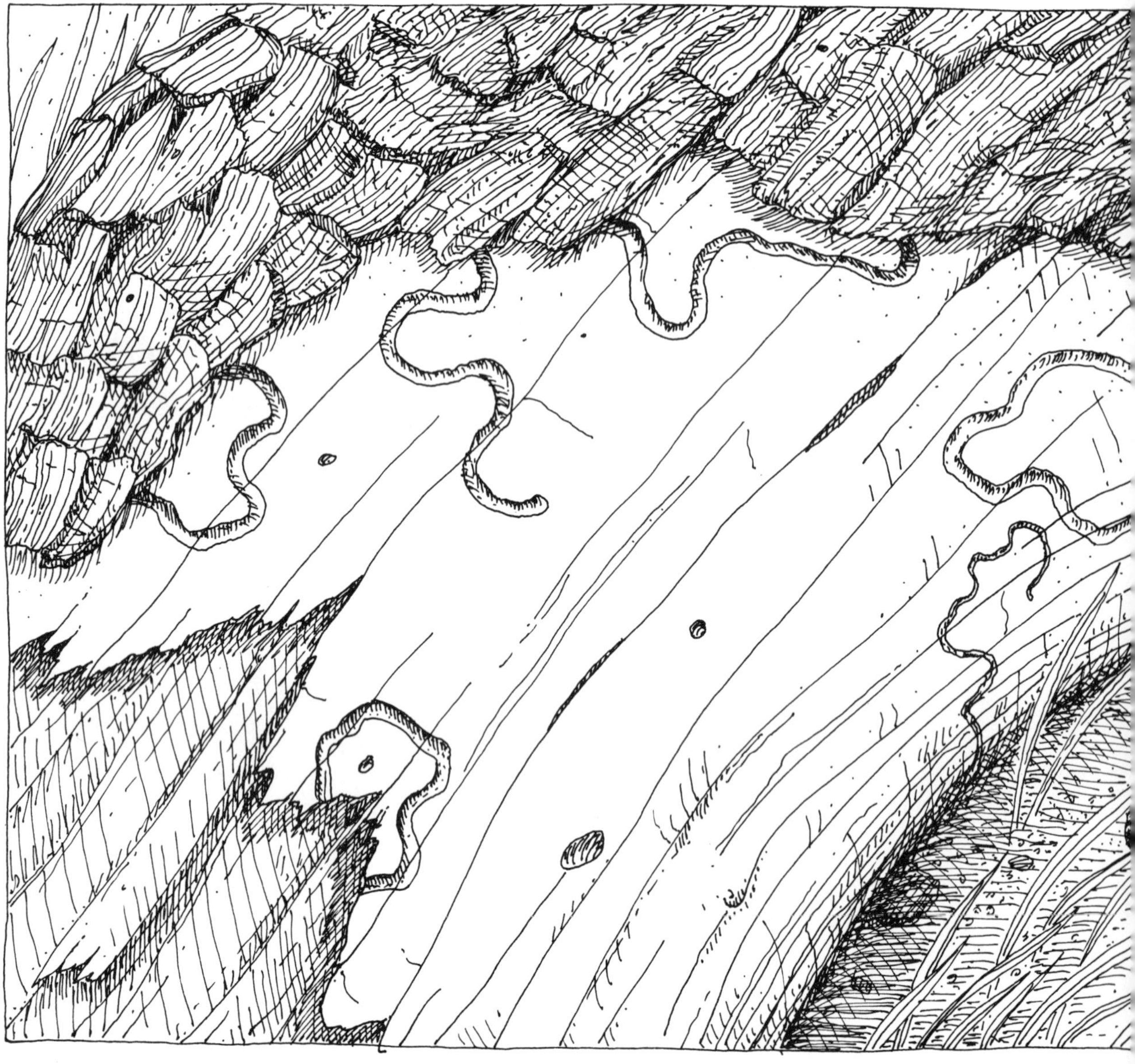

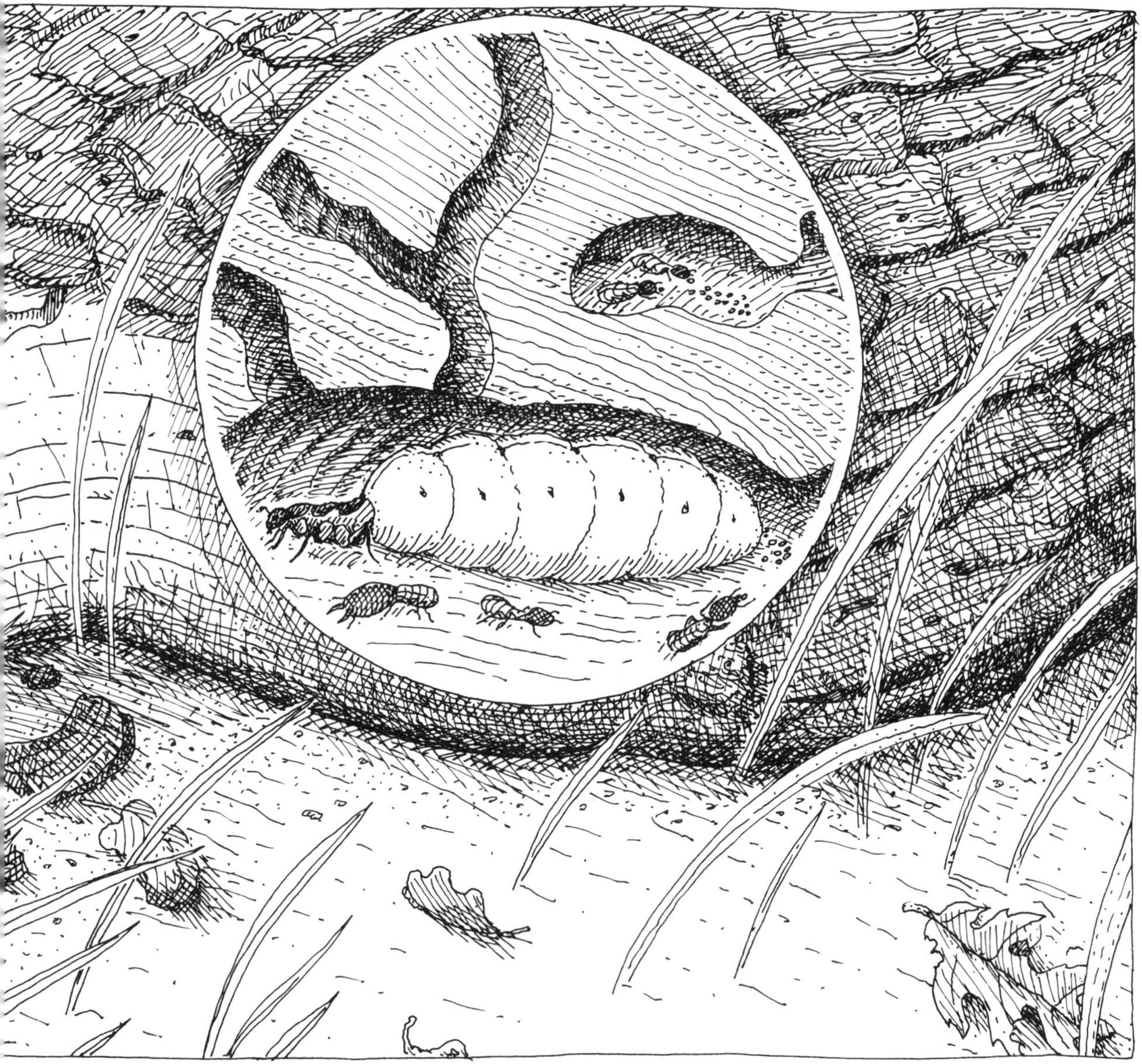

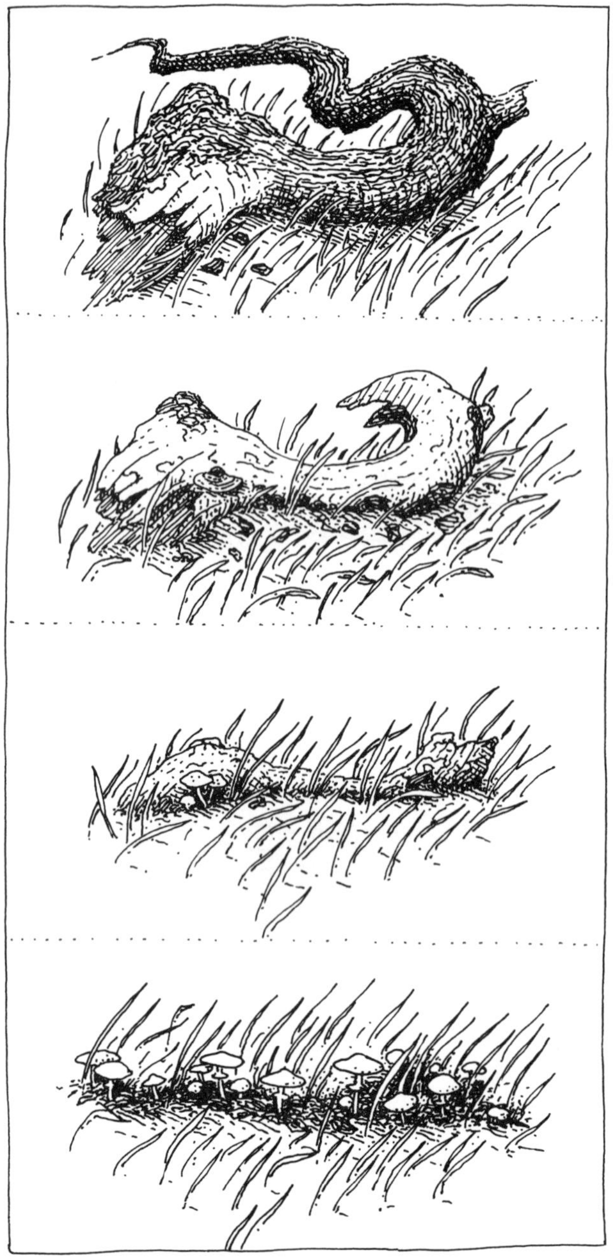

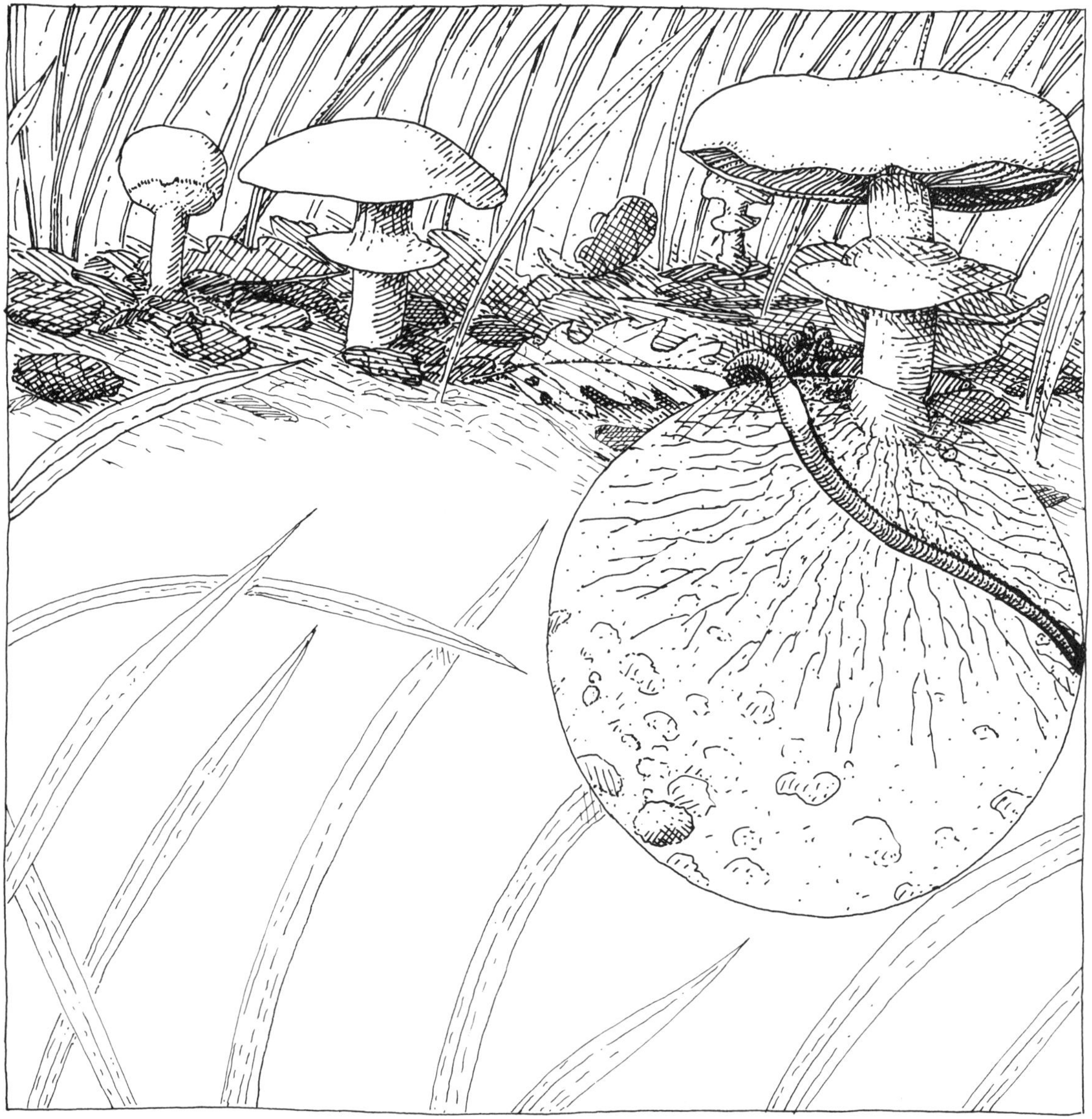

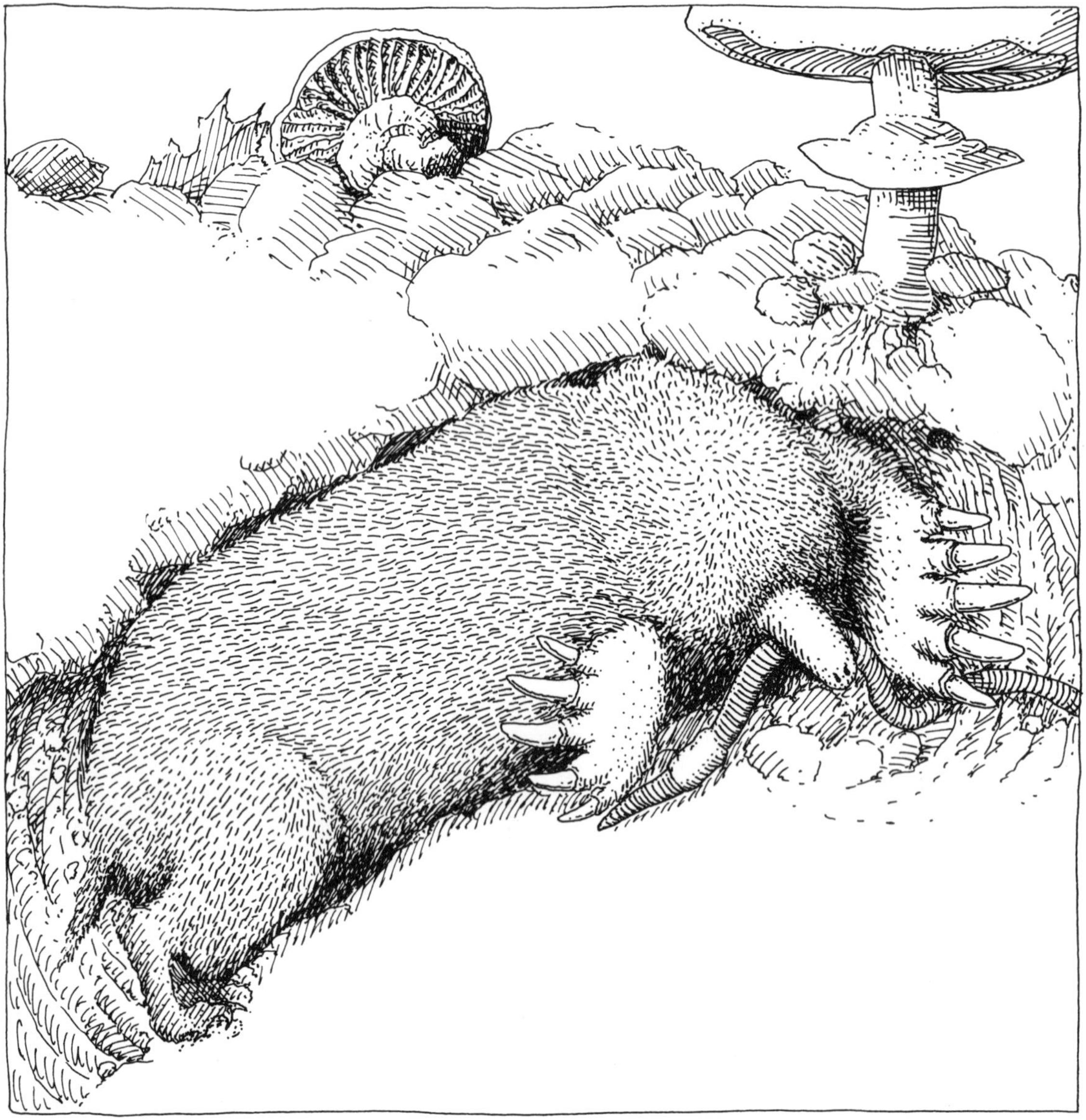

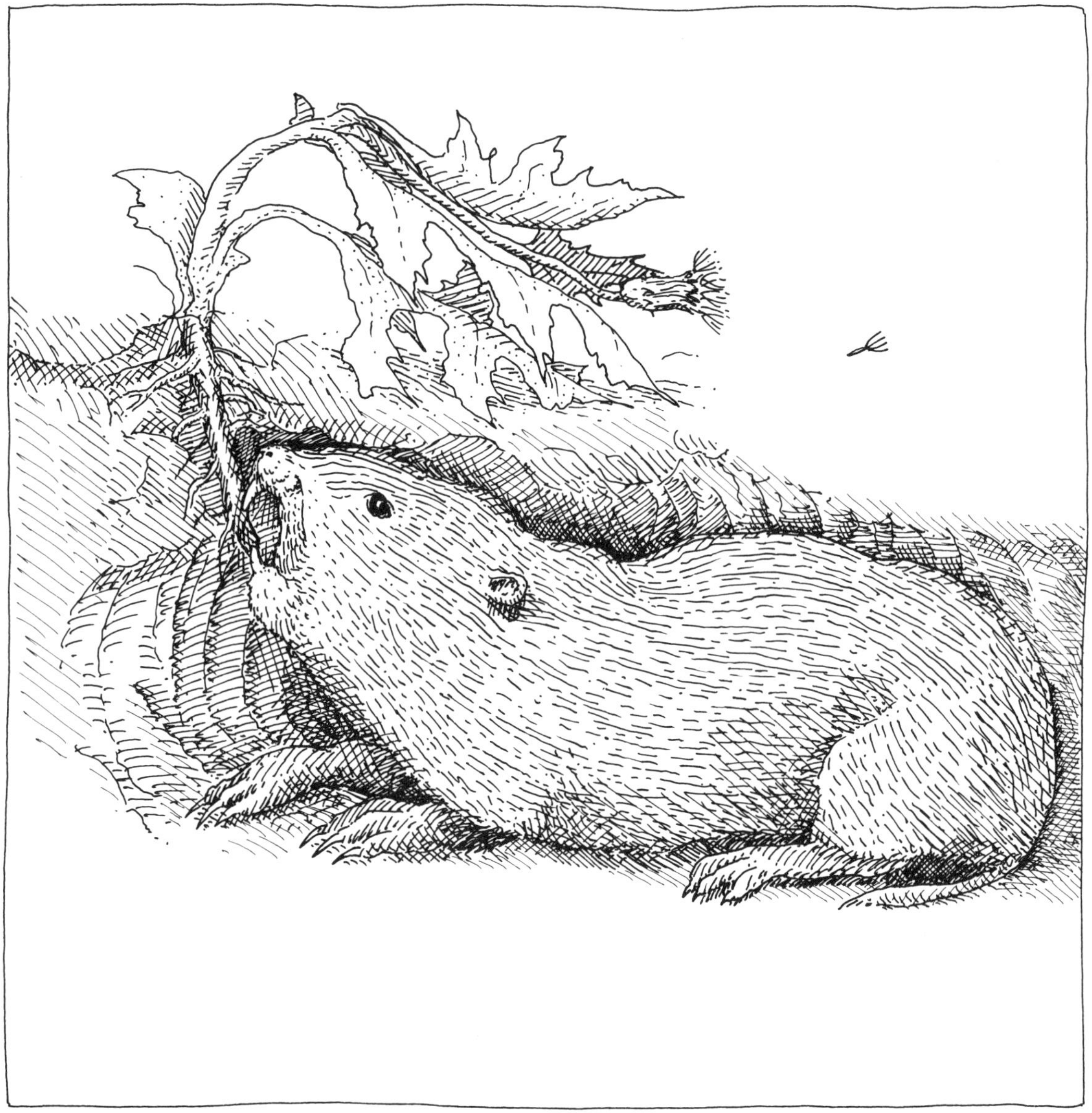

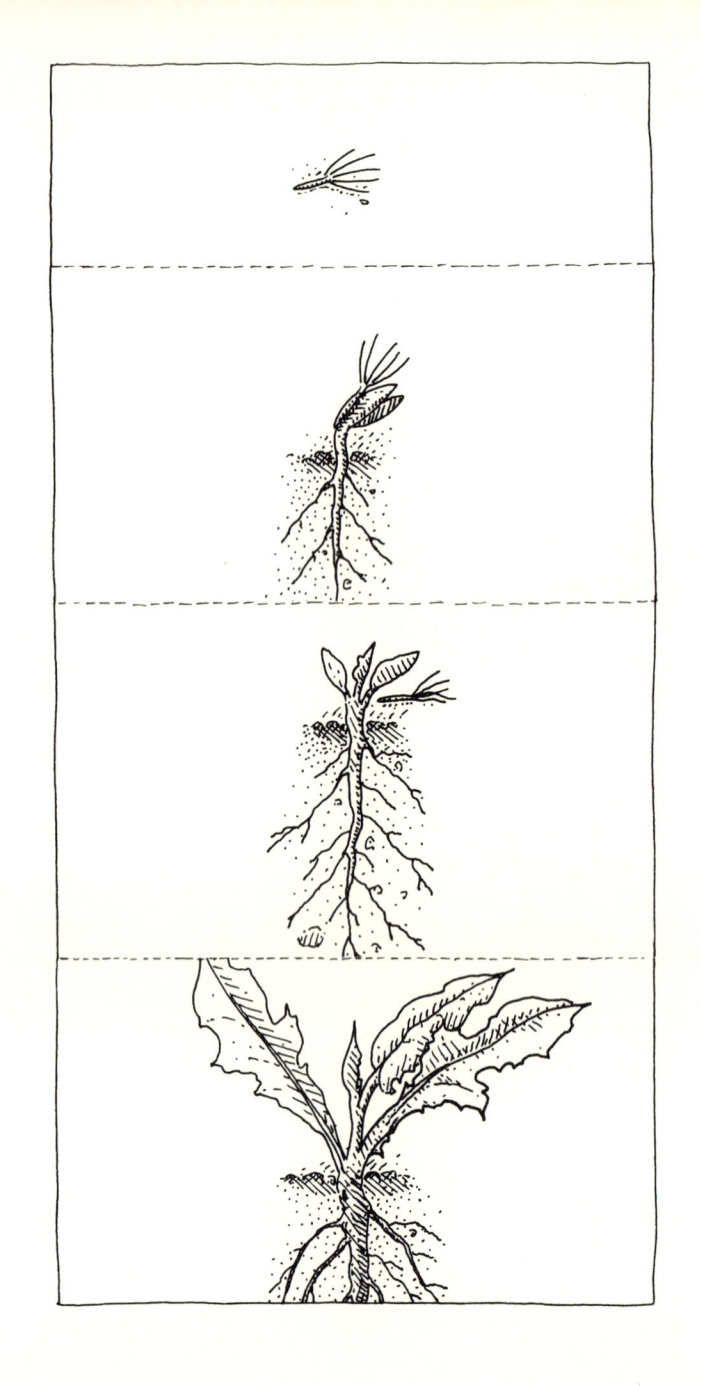

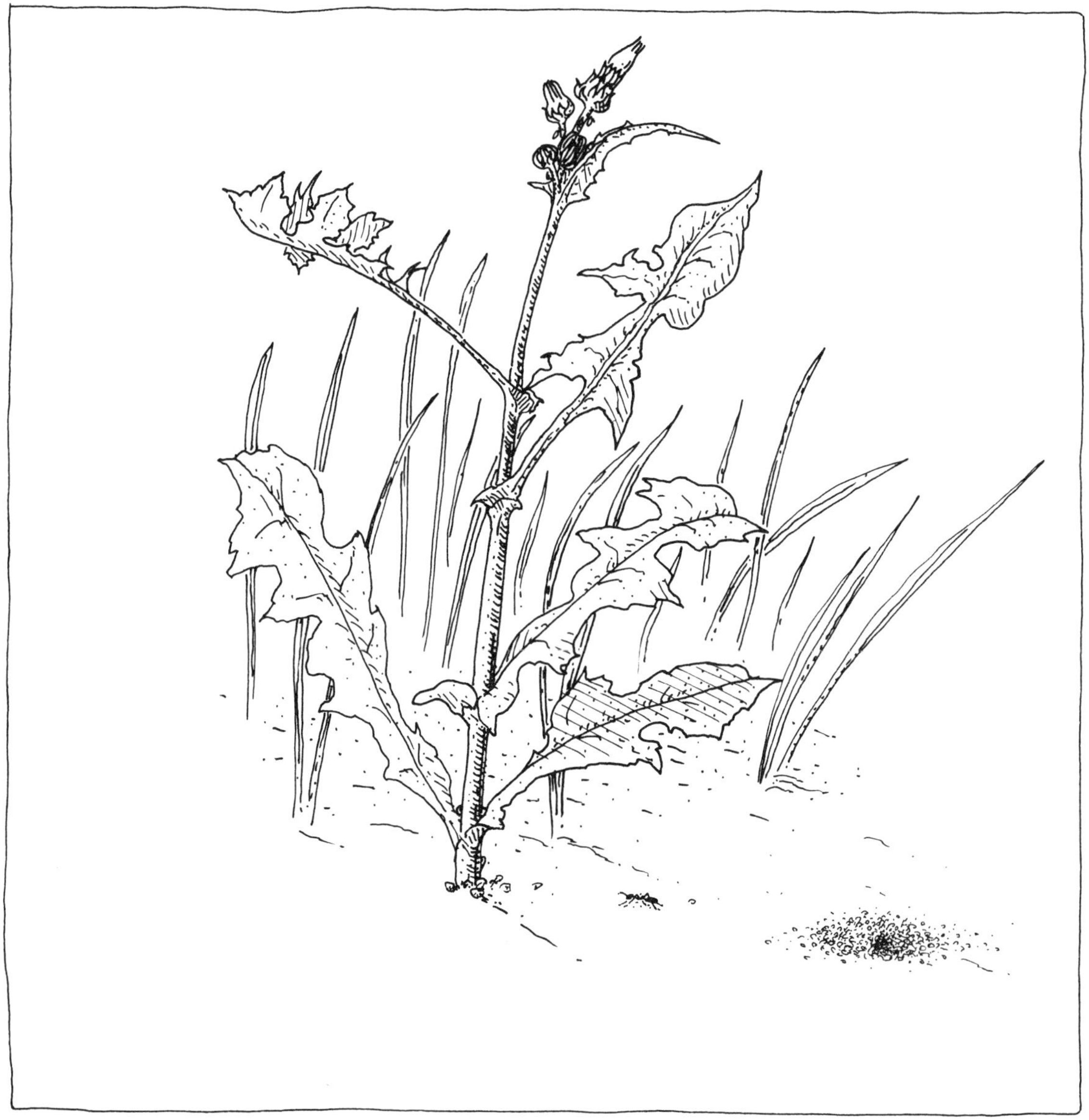

# III

# THE GREEN LEAF CREATURES

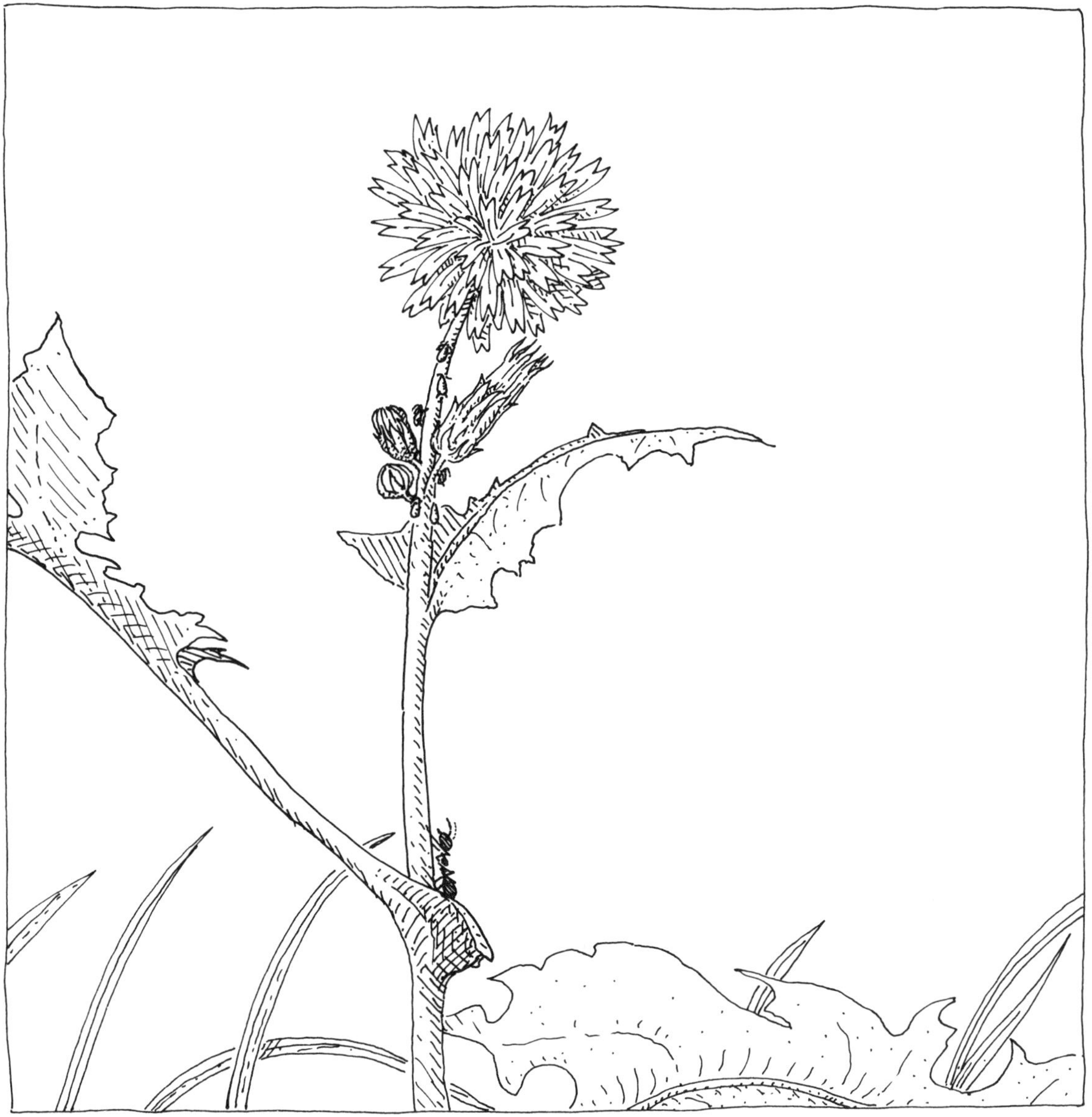

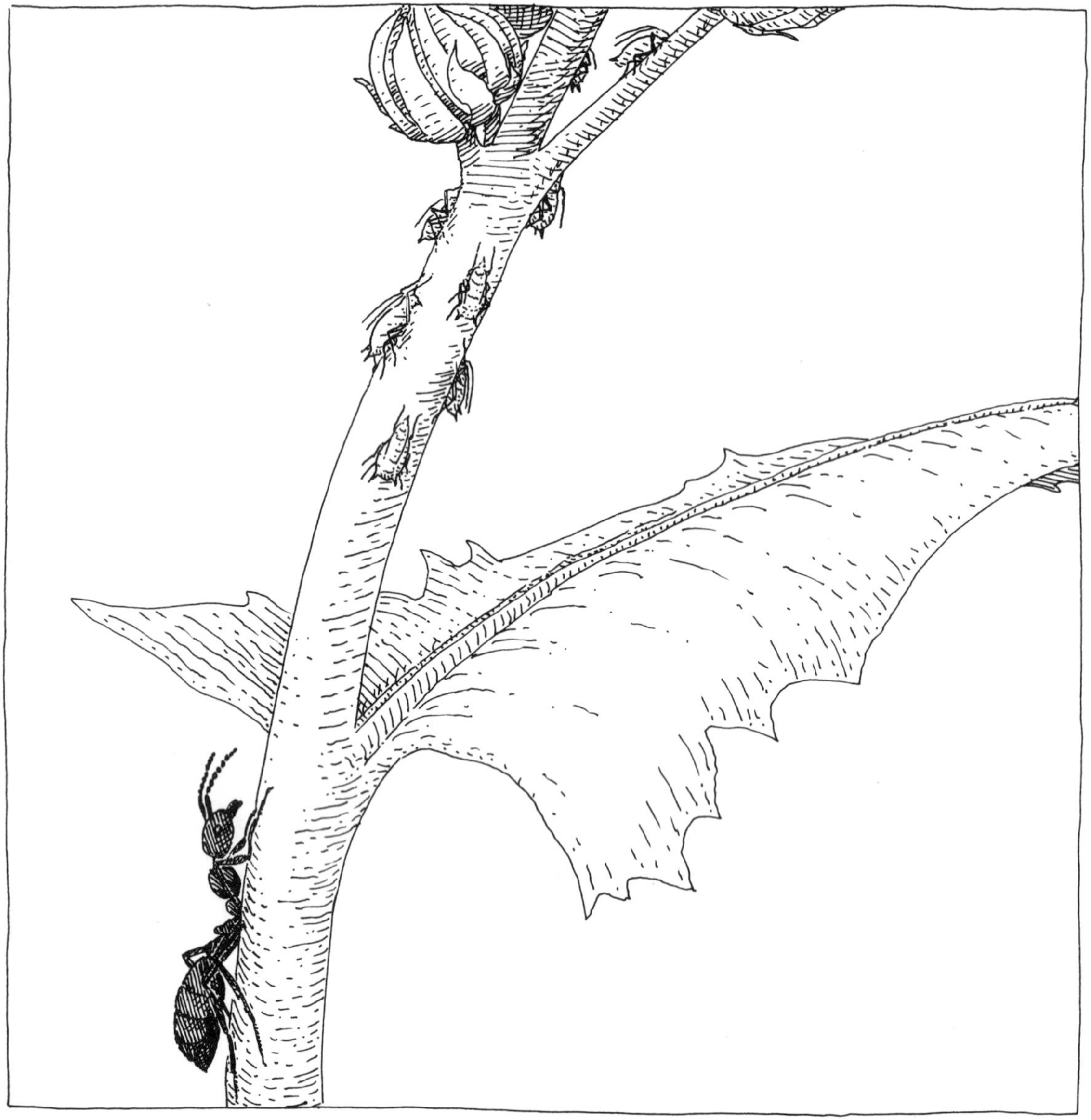

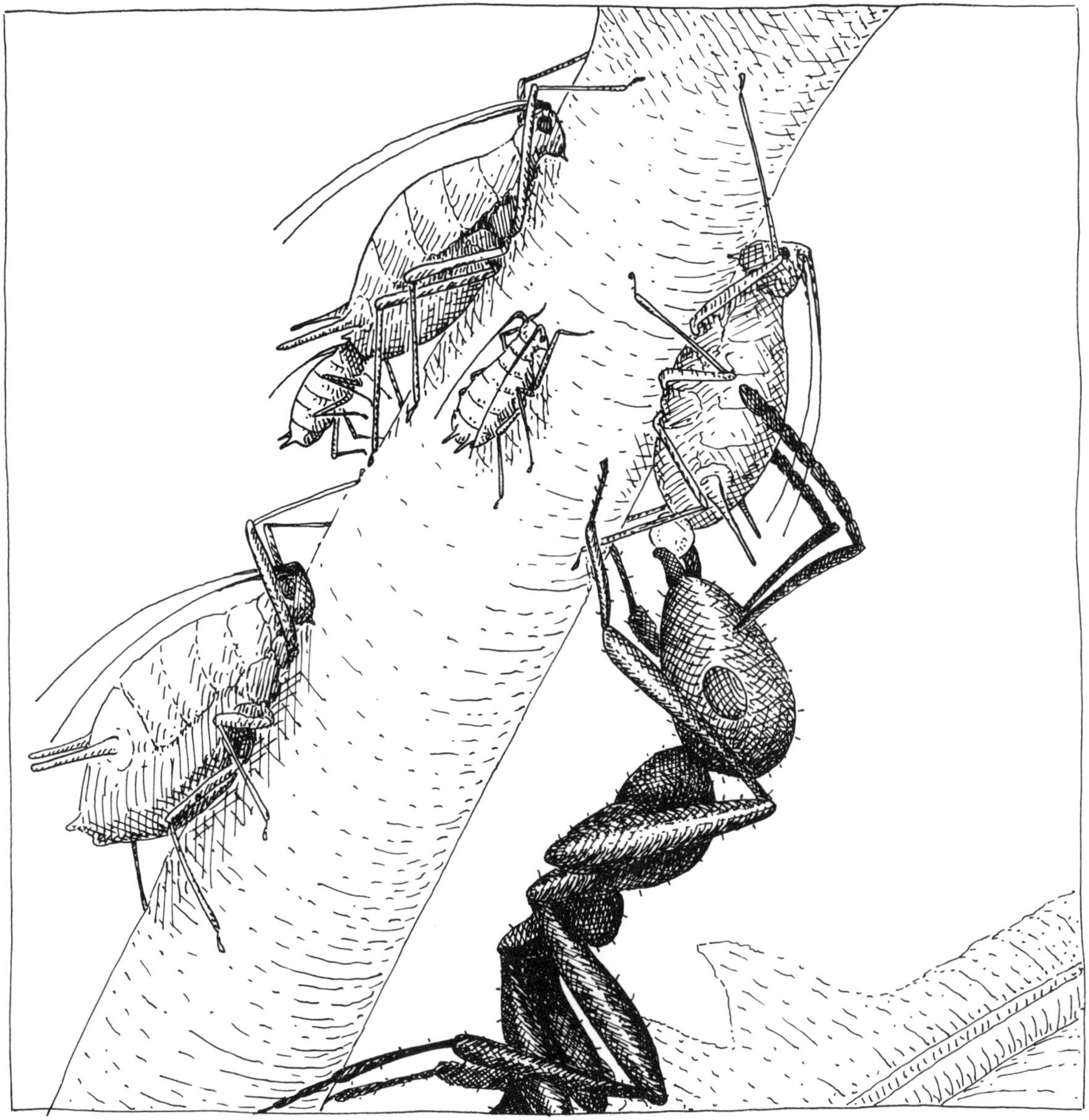

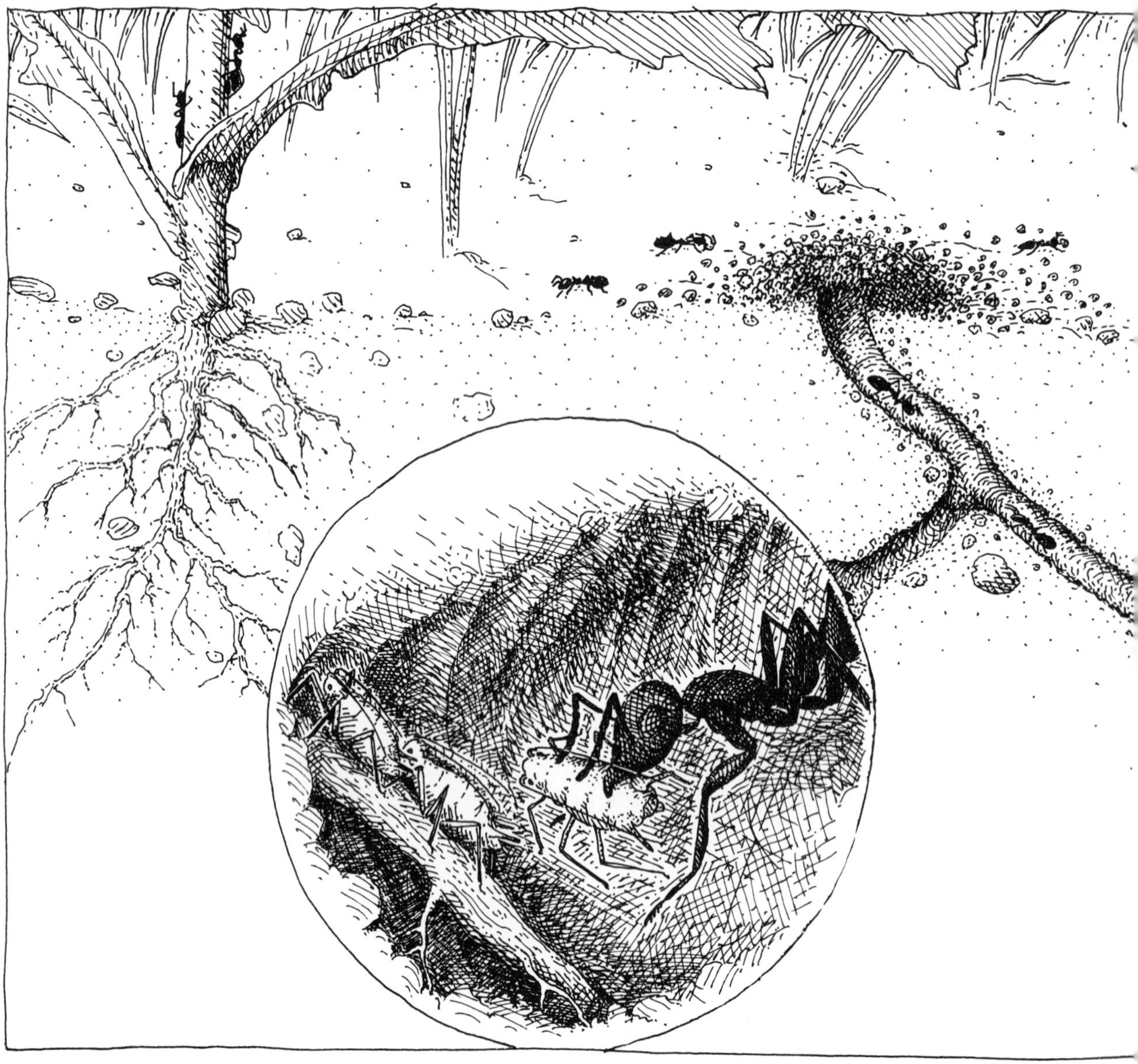

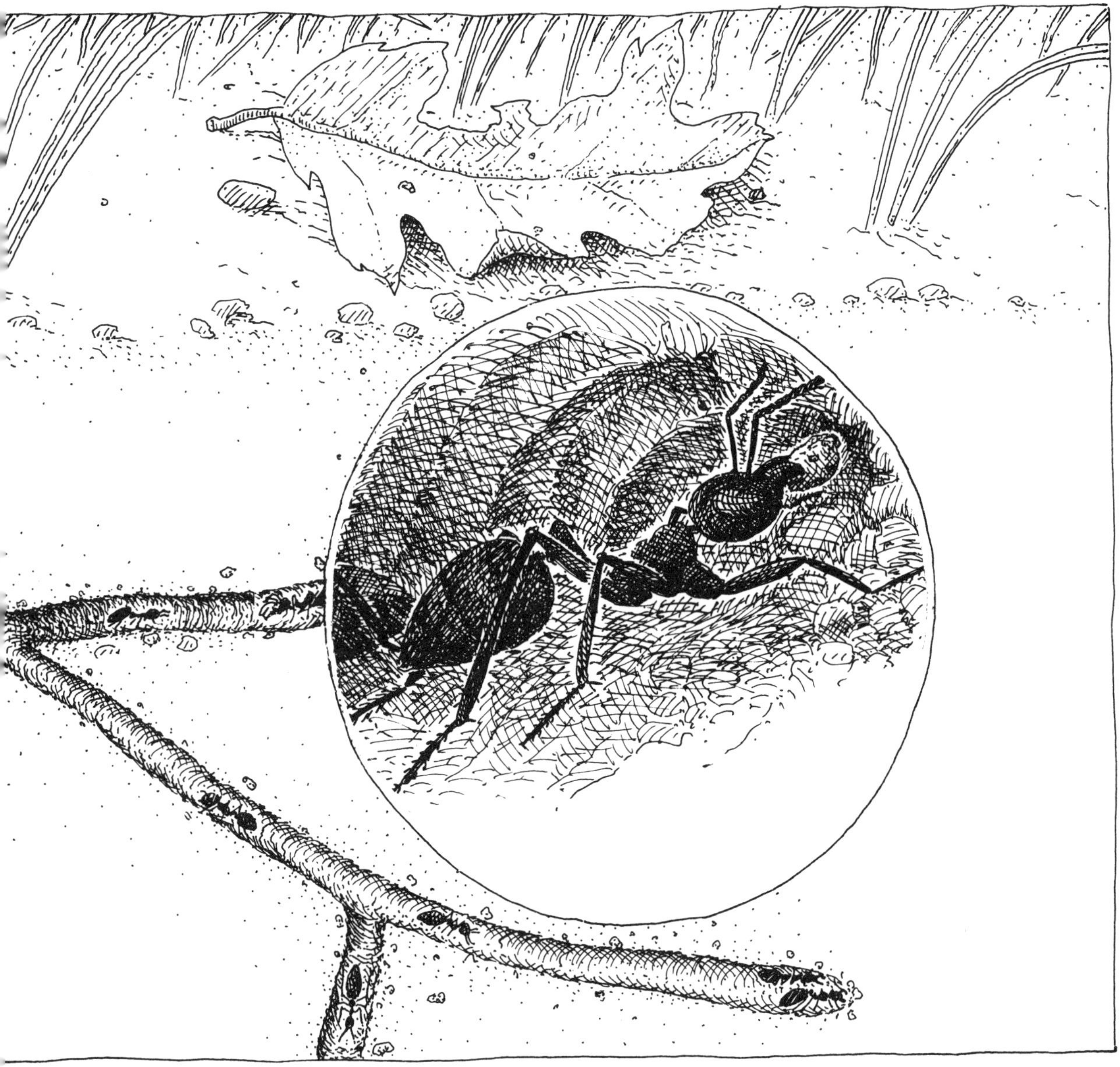

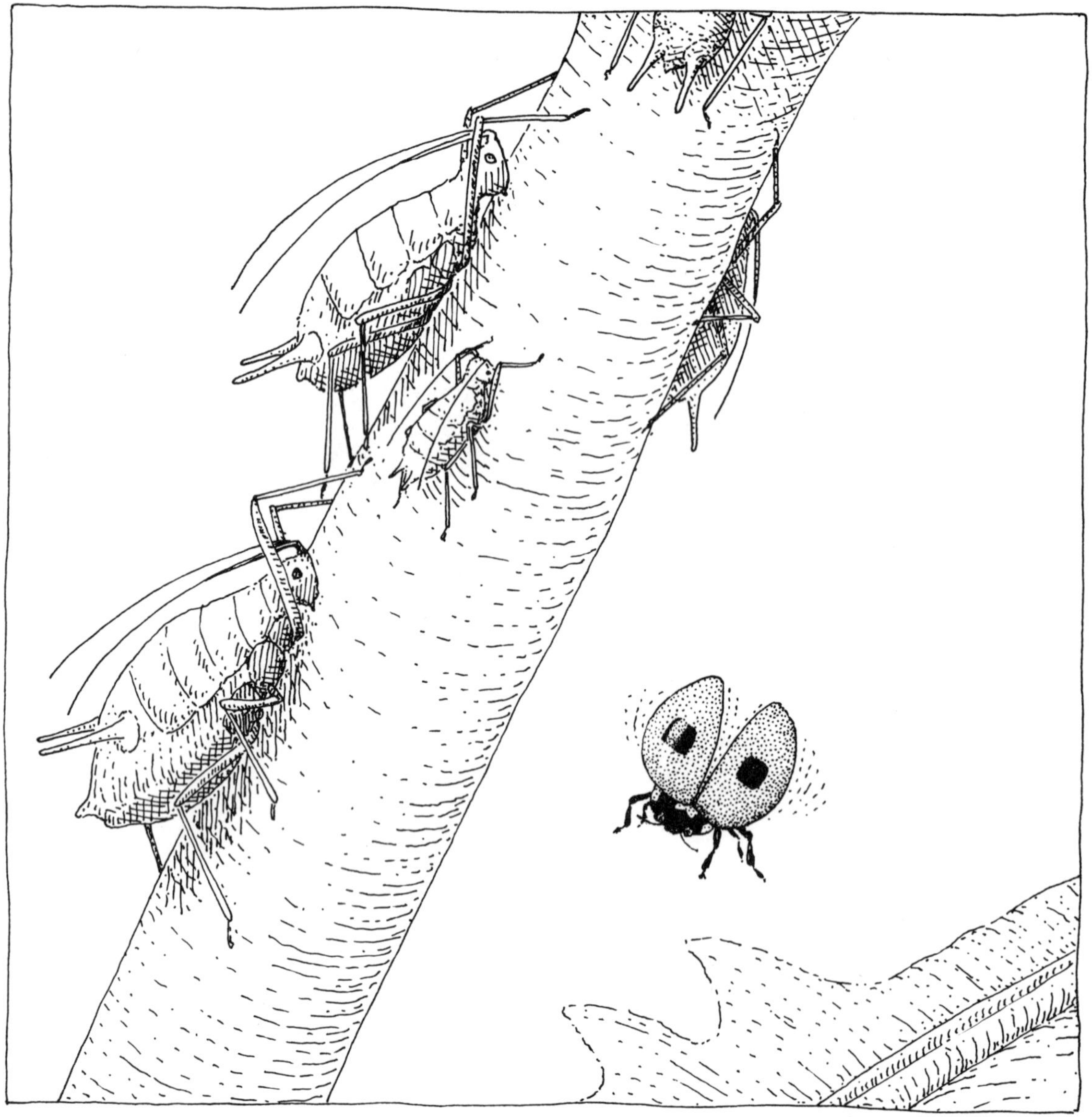

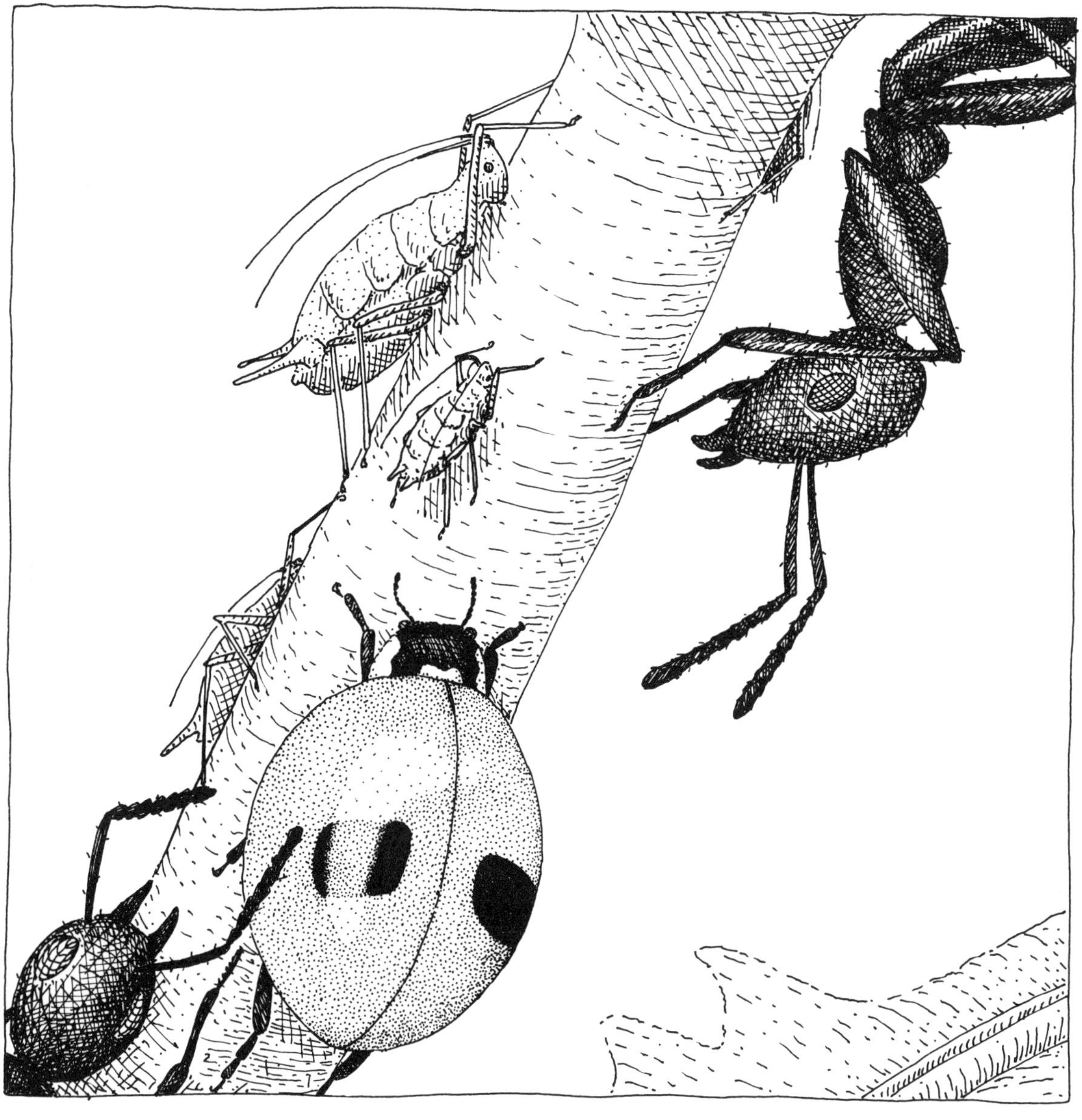

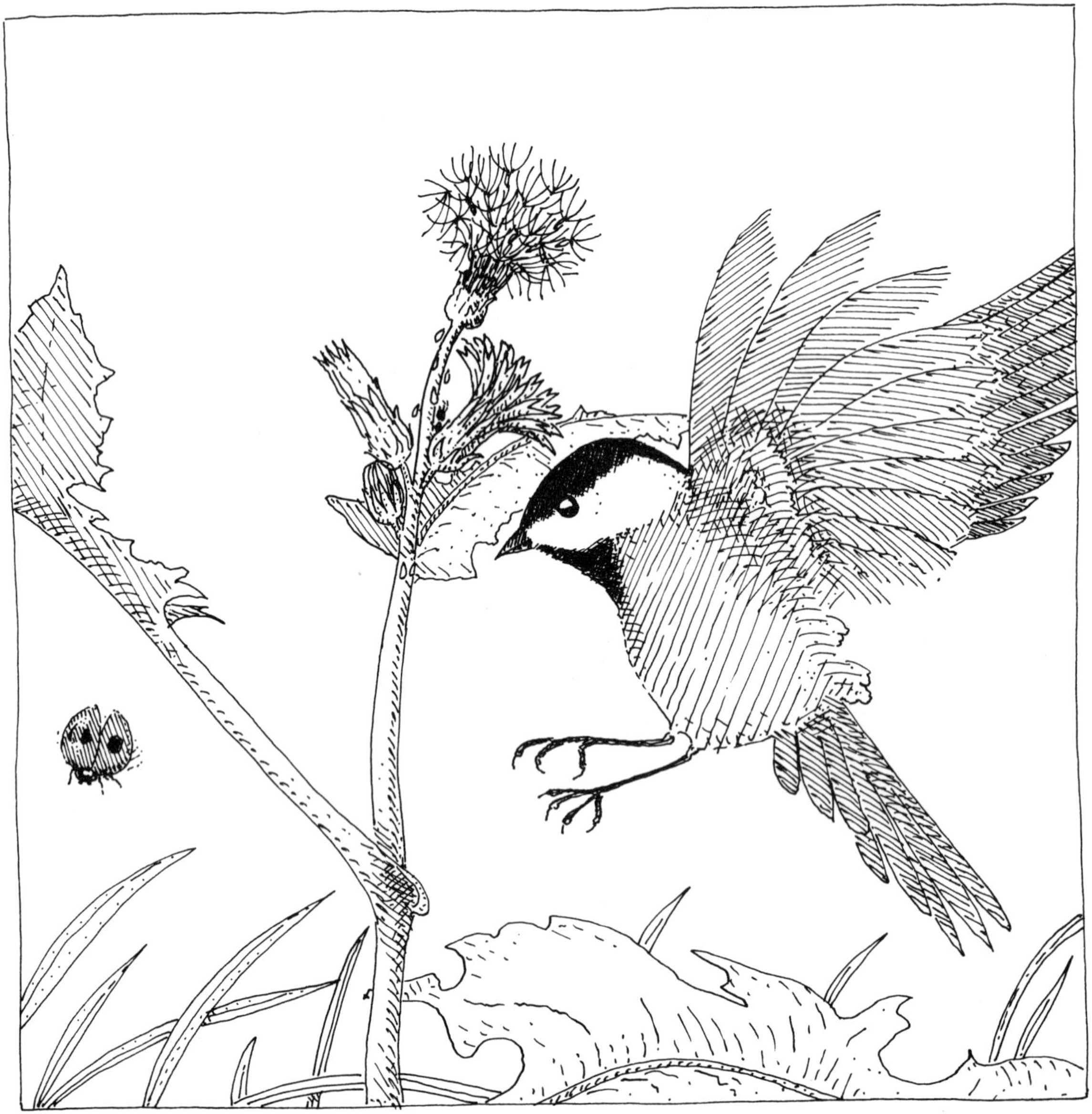

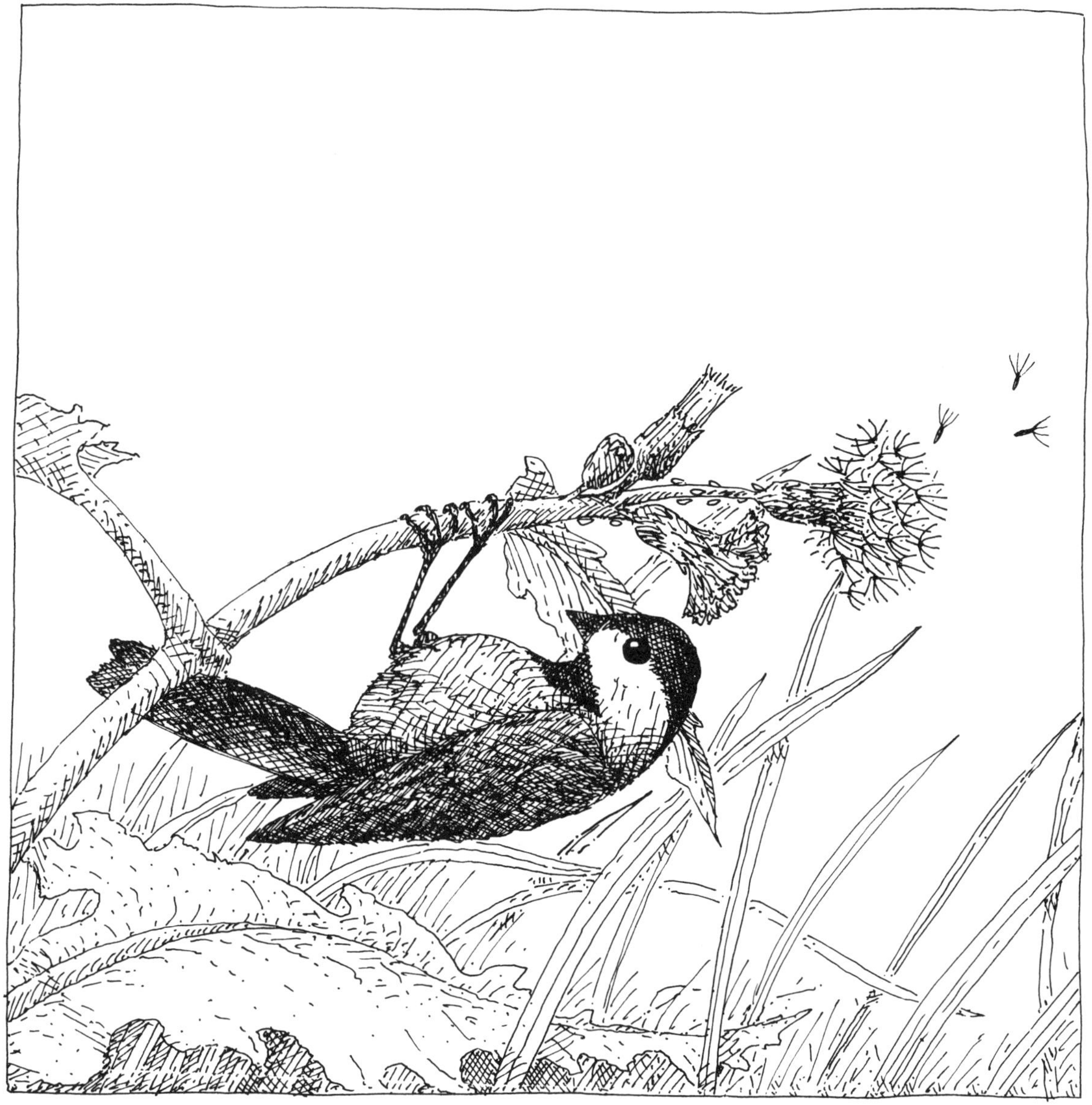

# IV
# A TURTLE'S WORLD

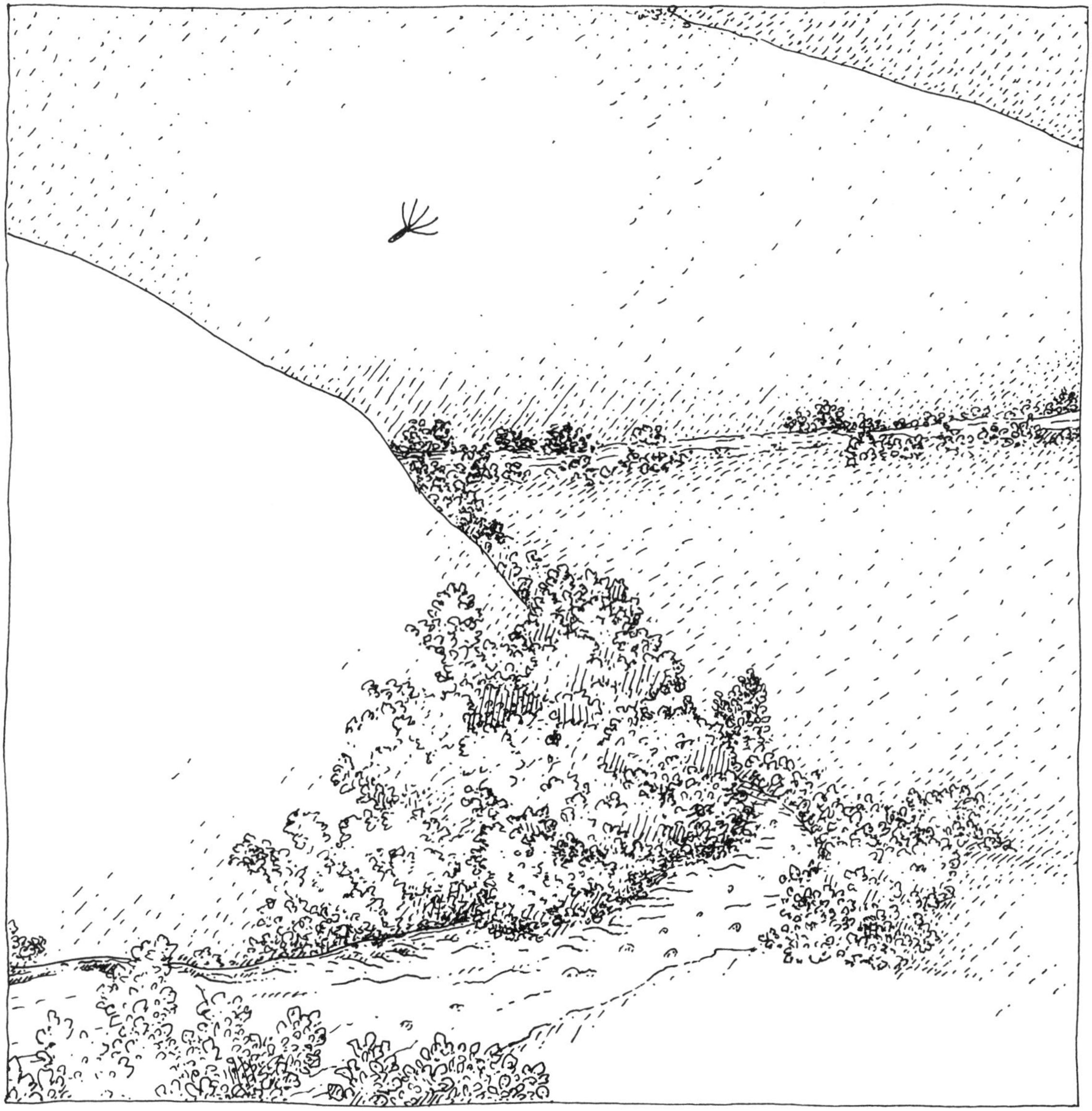

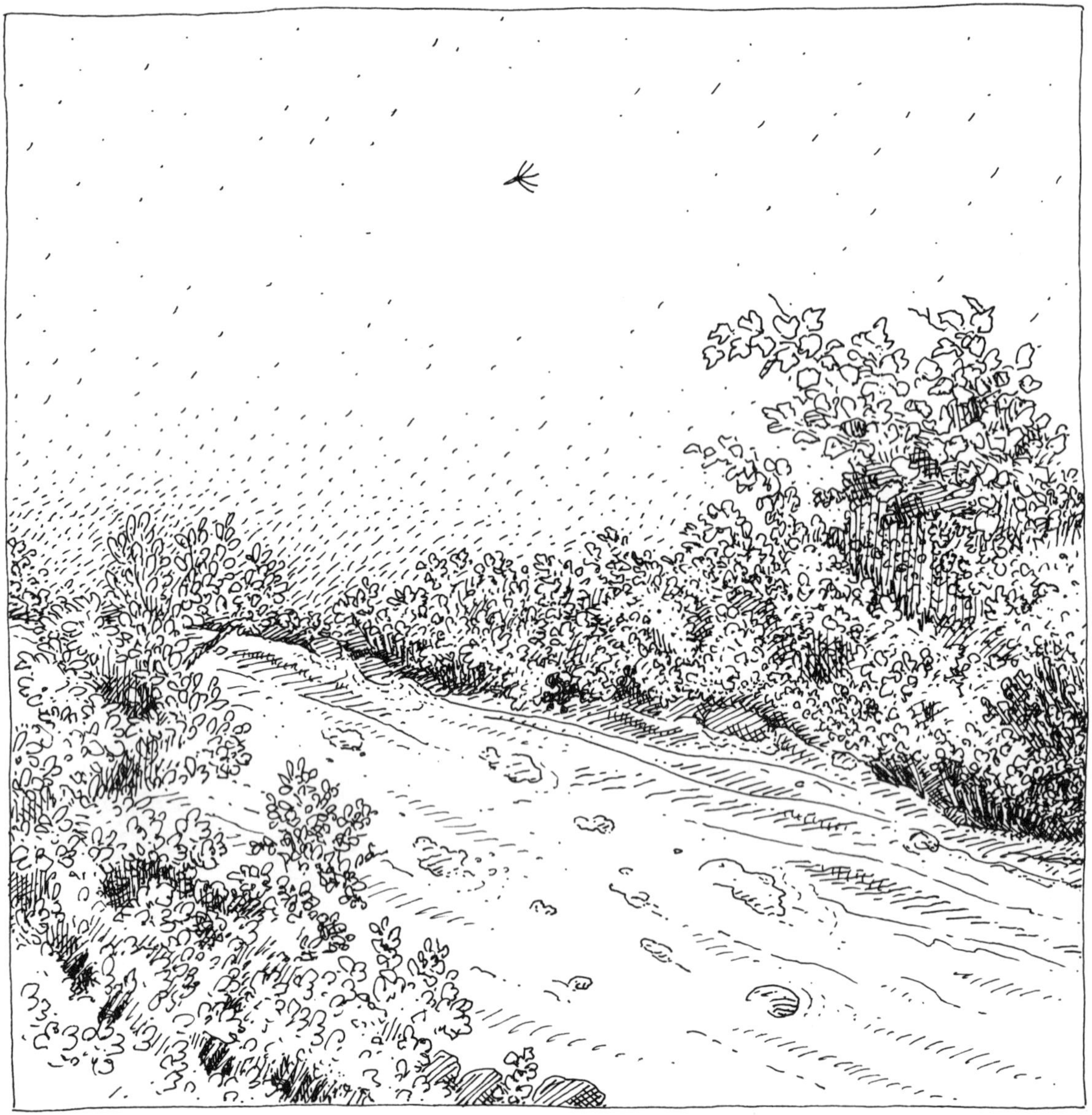

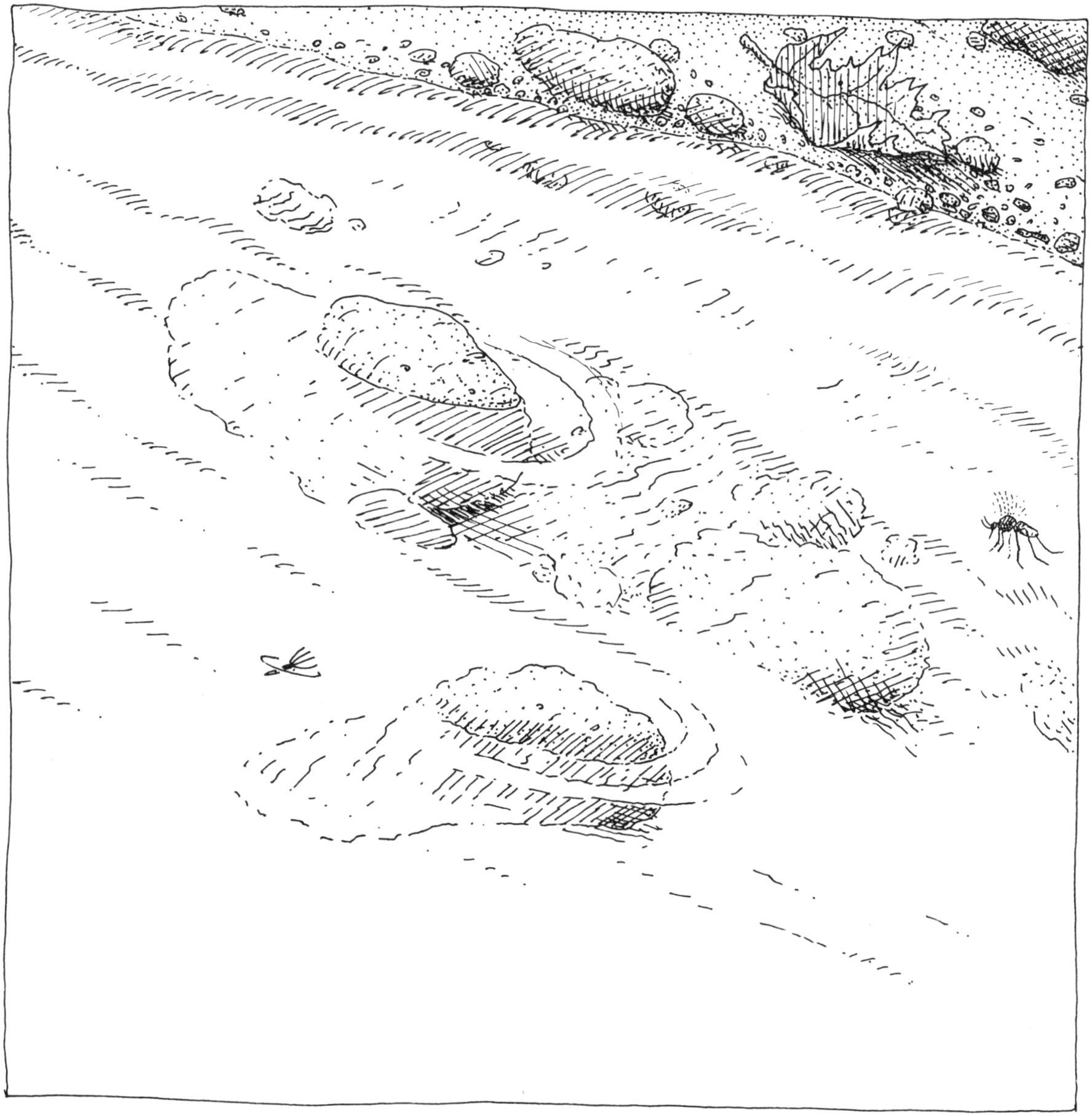

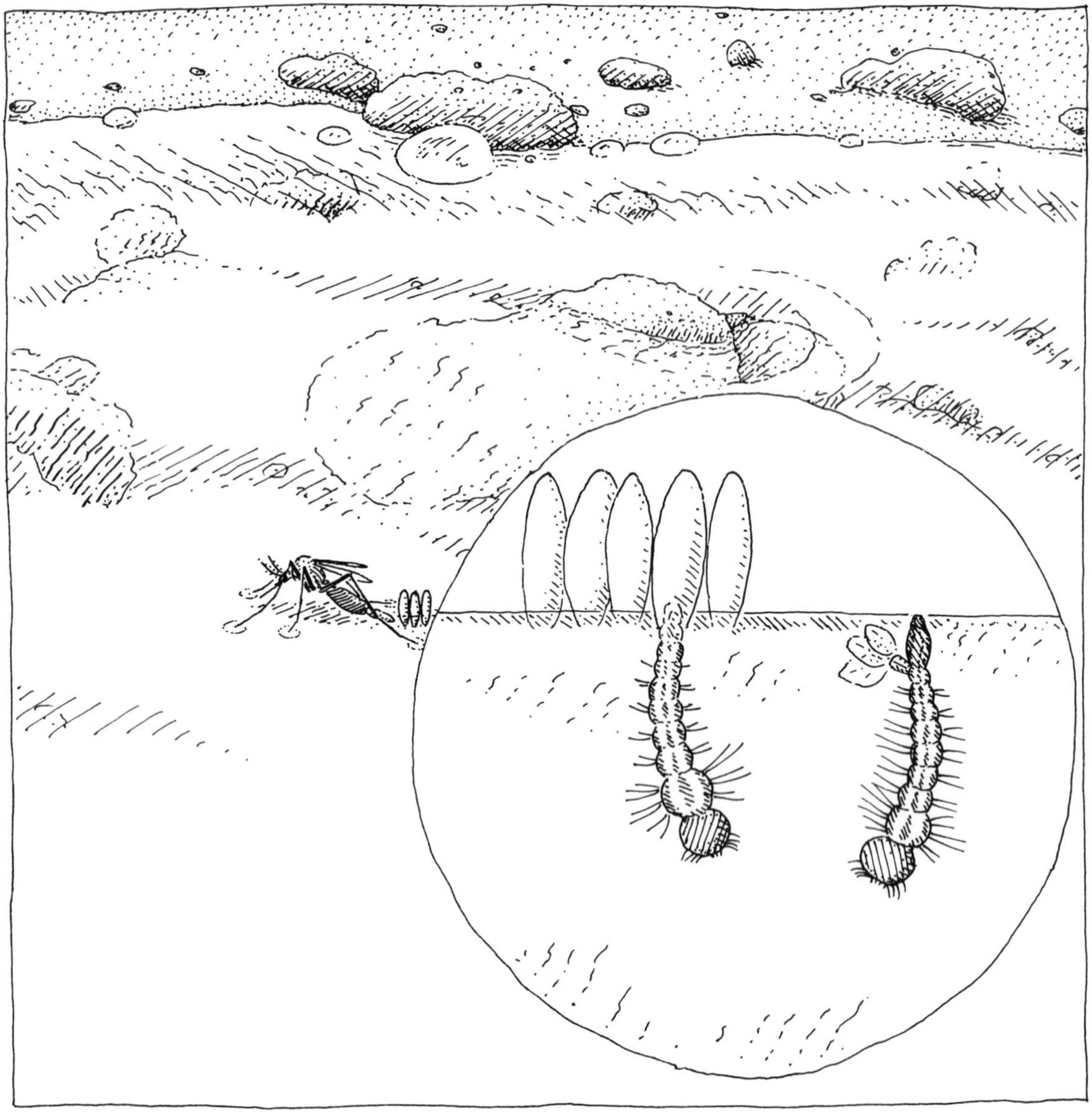

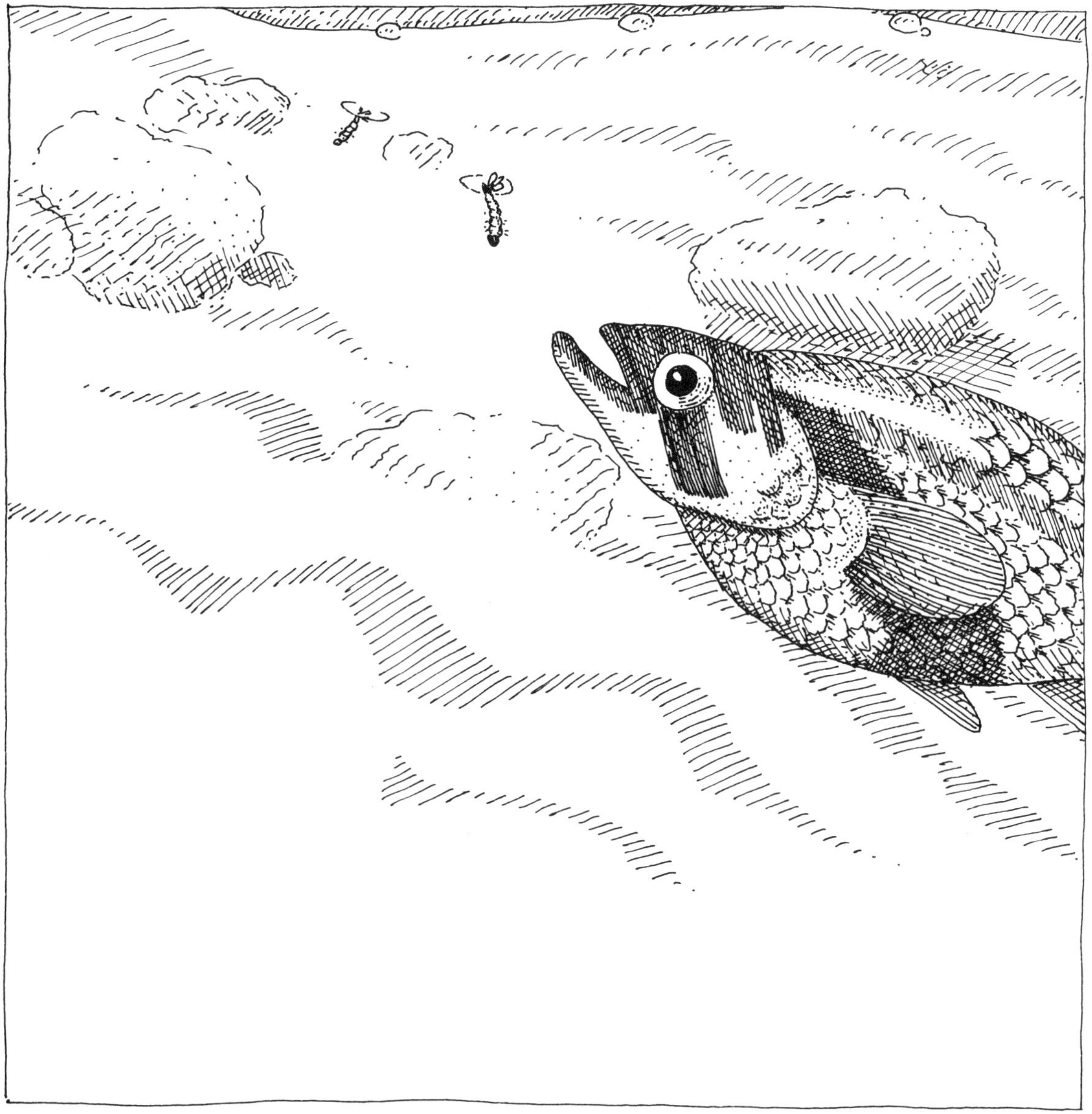

70

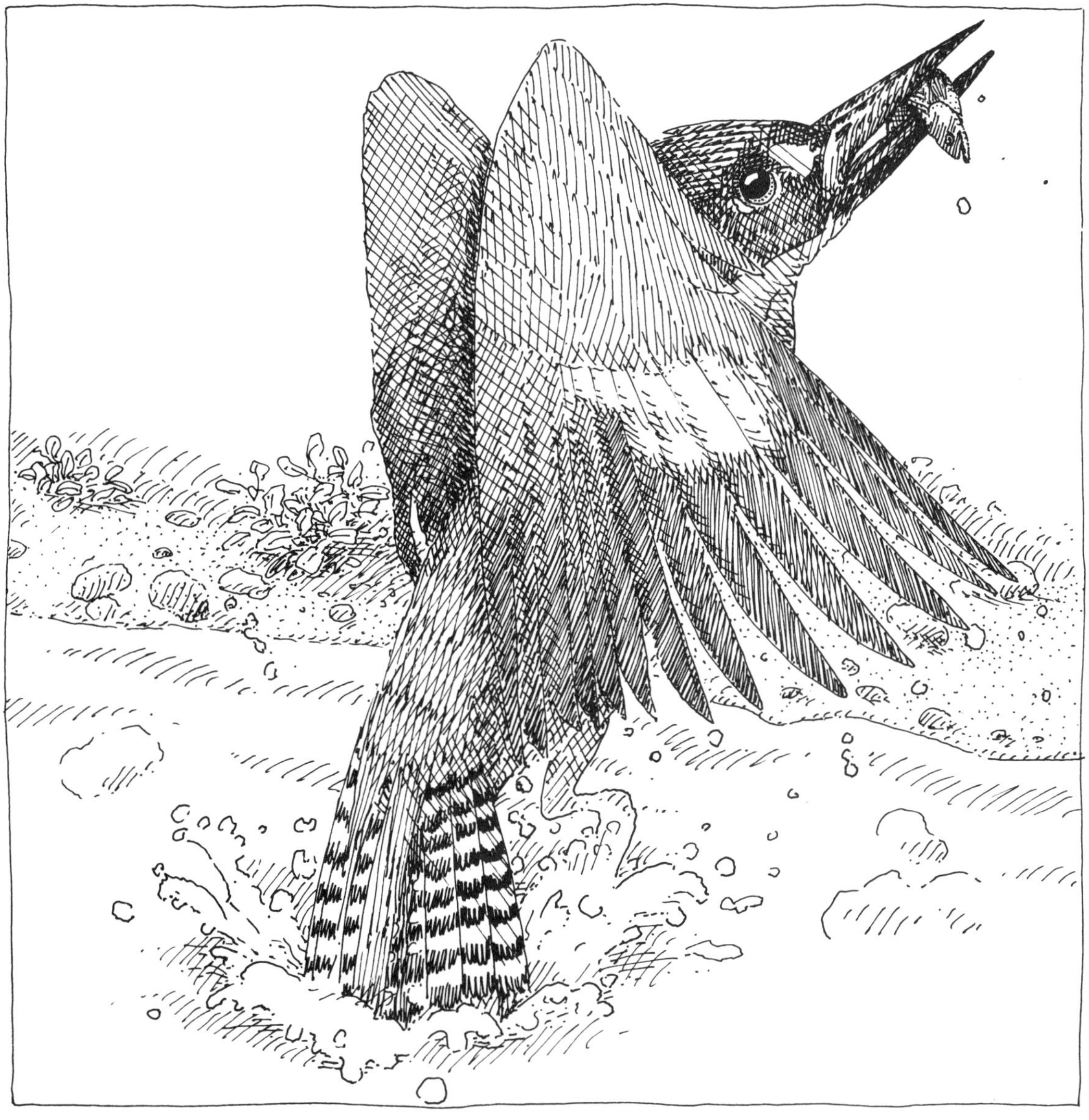

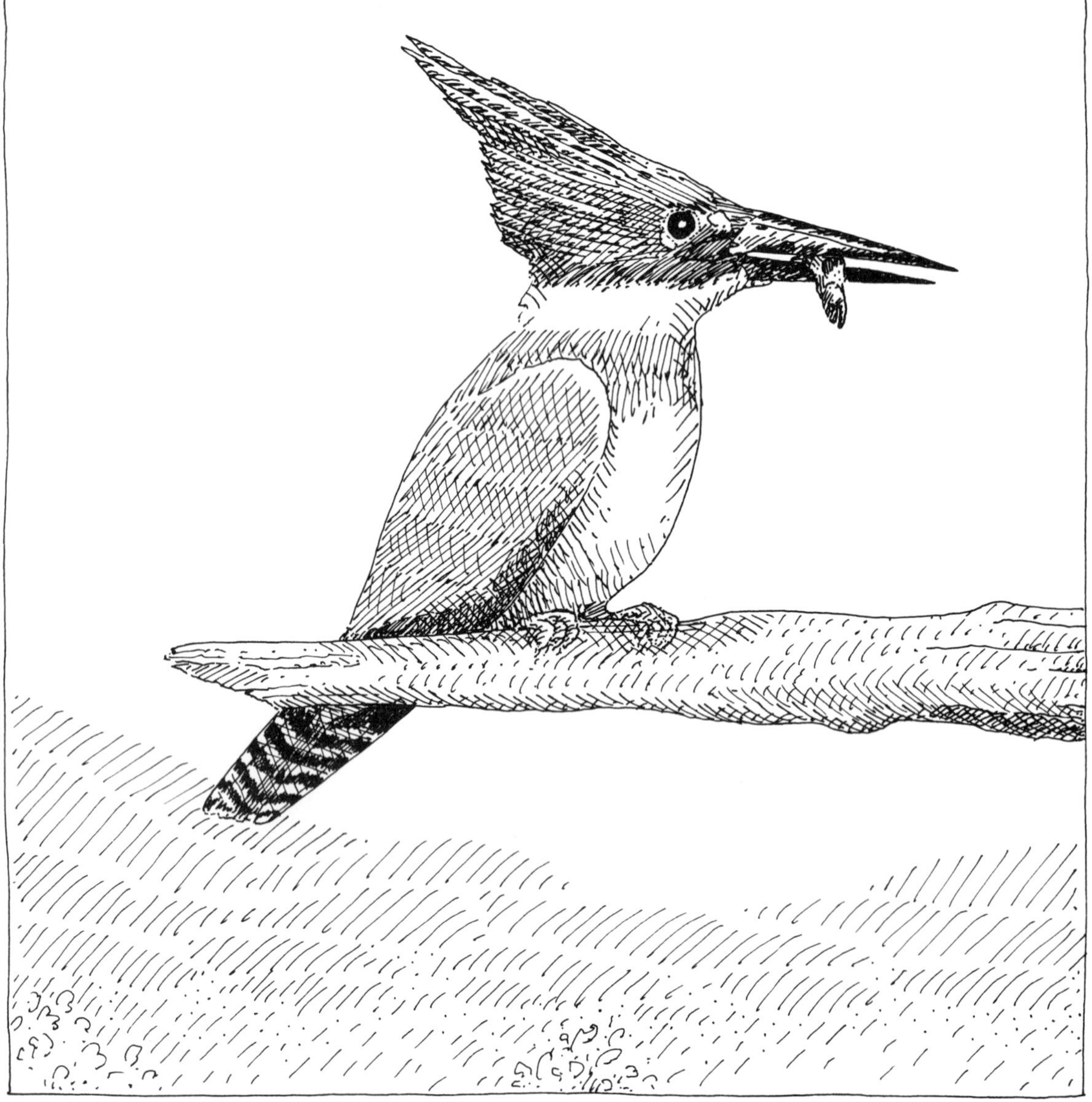

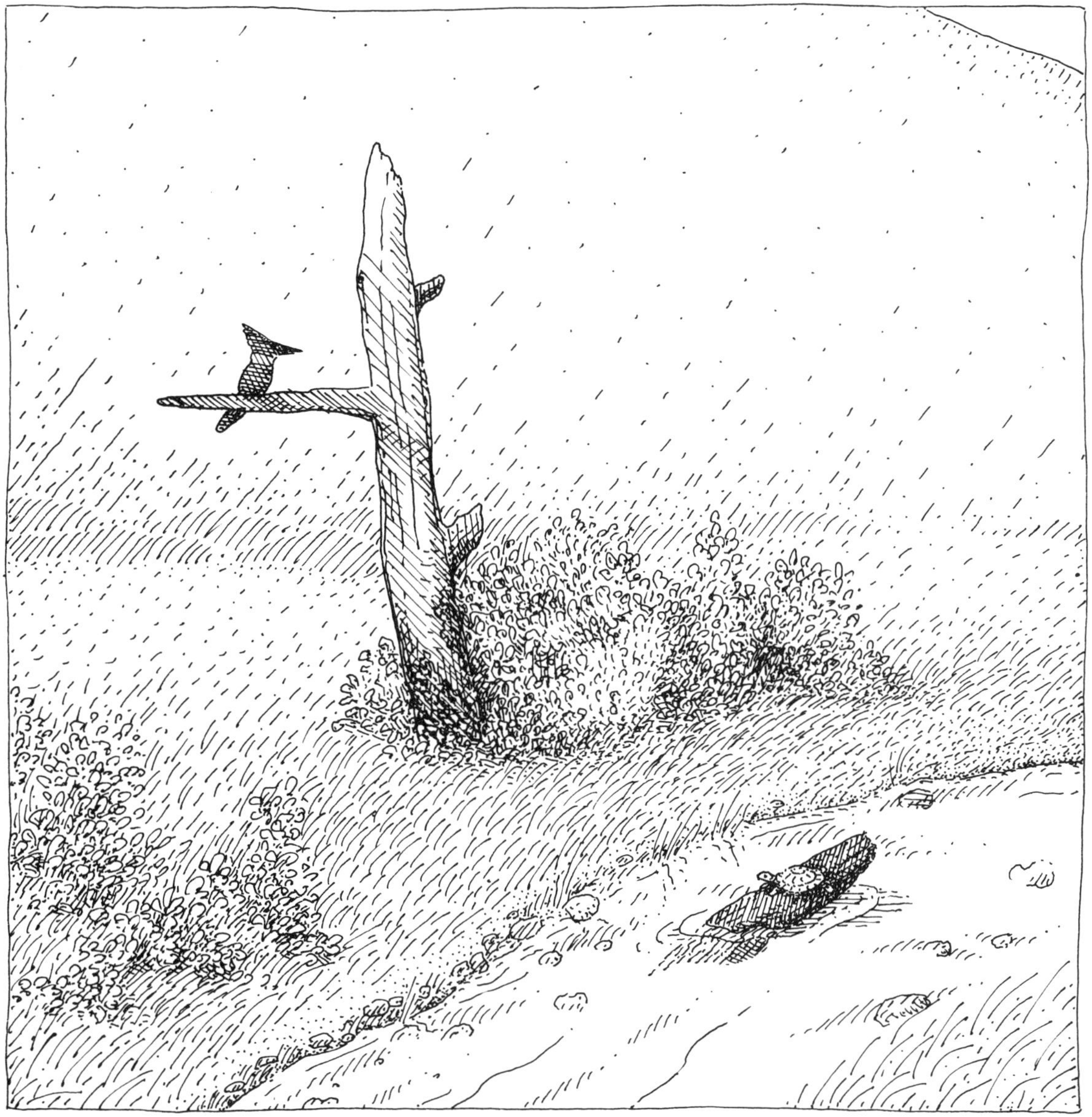

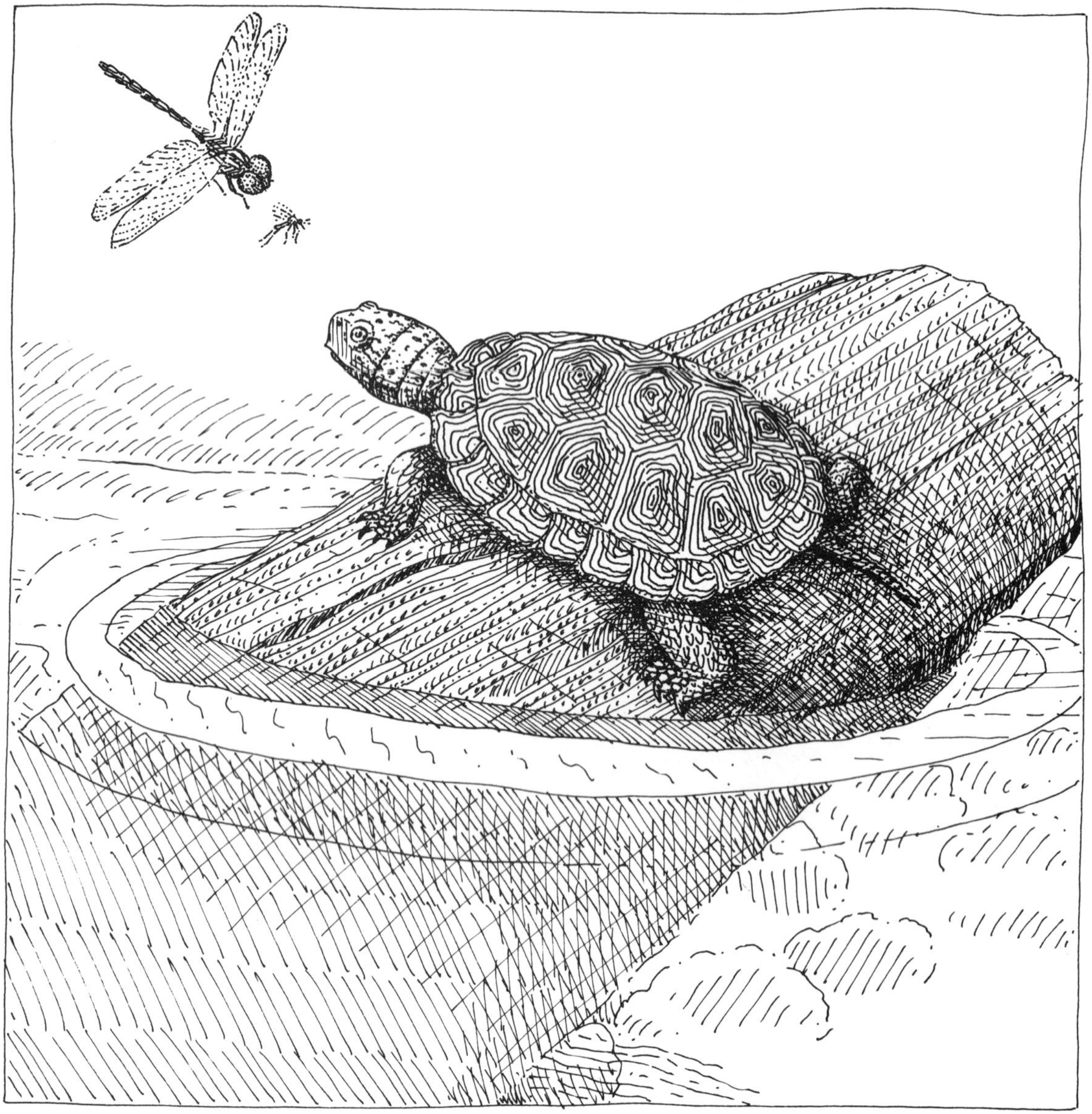

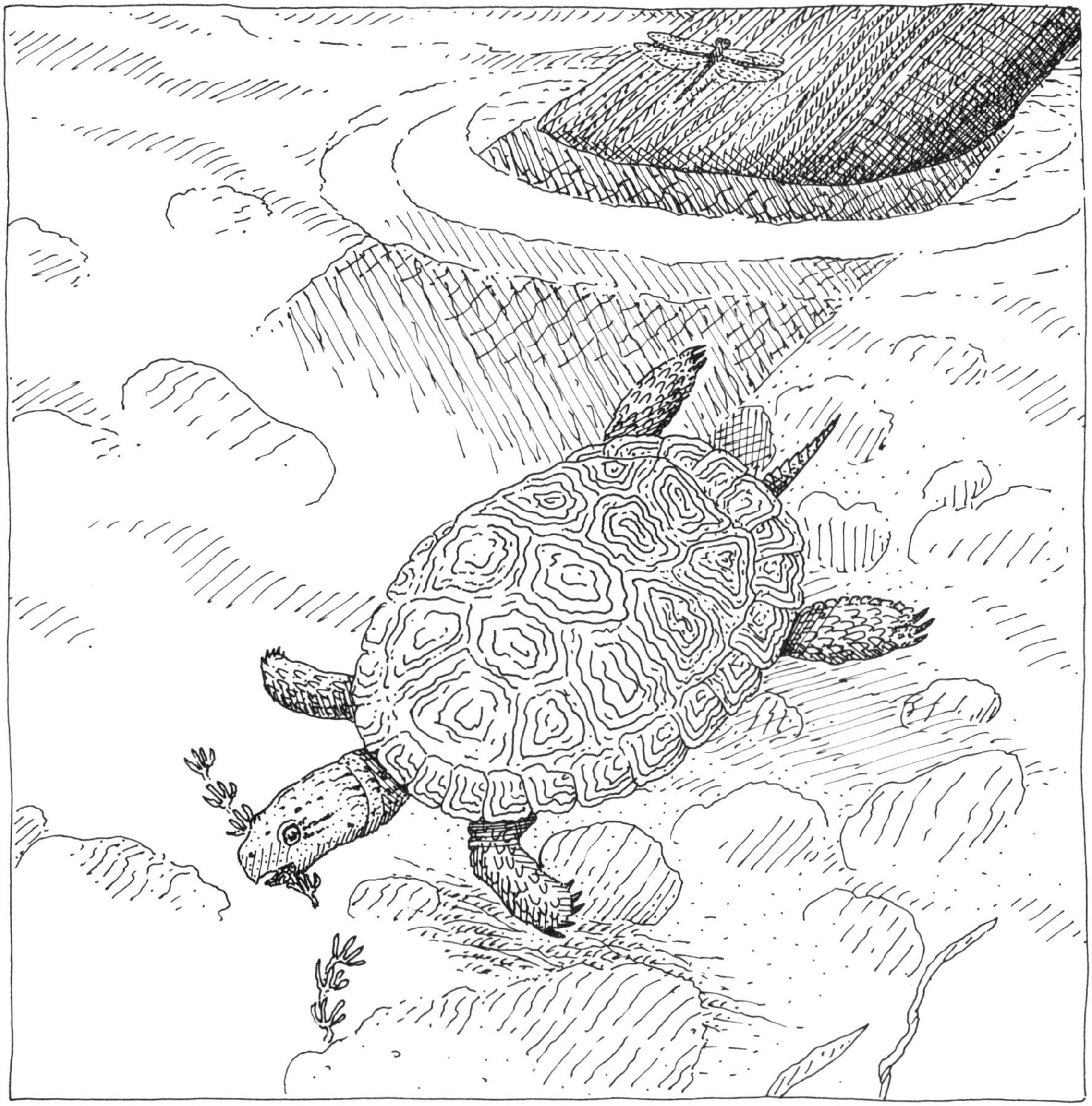

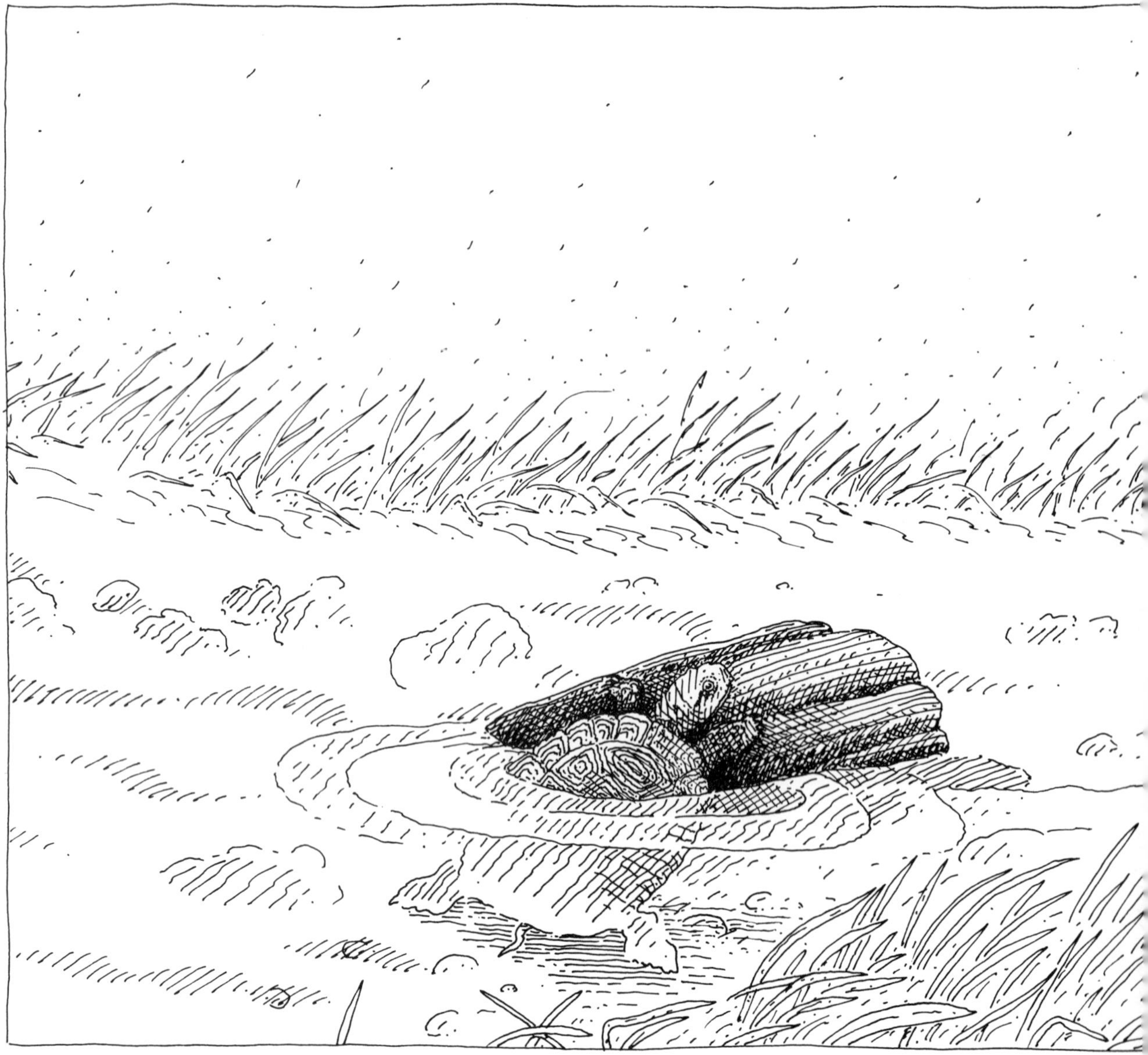

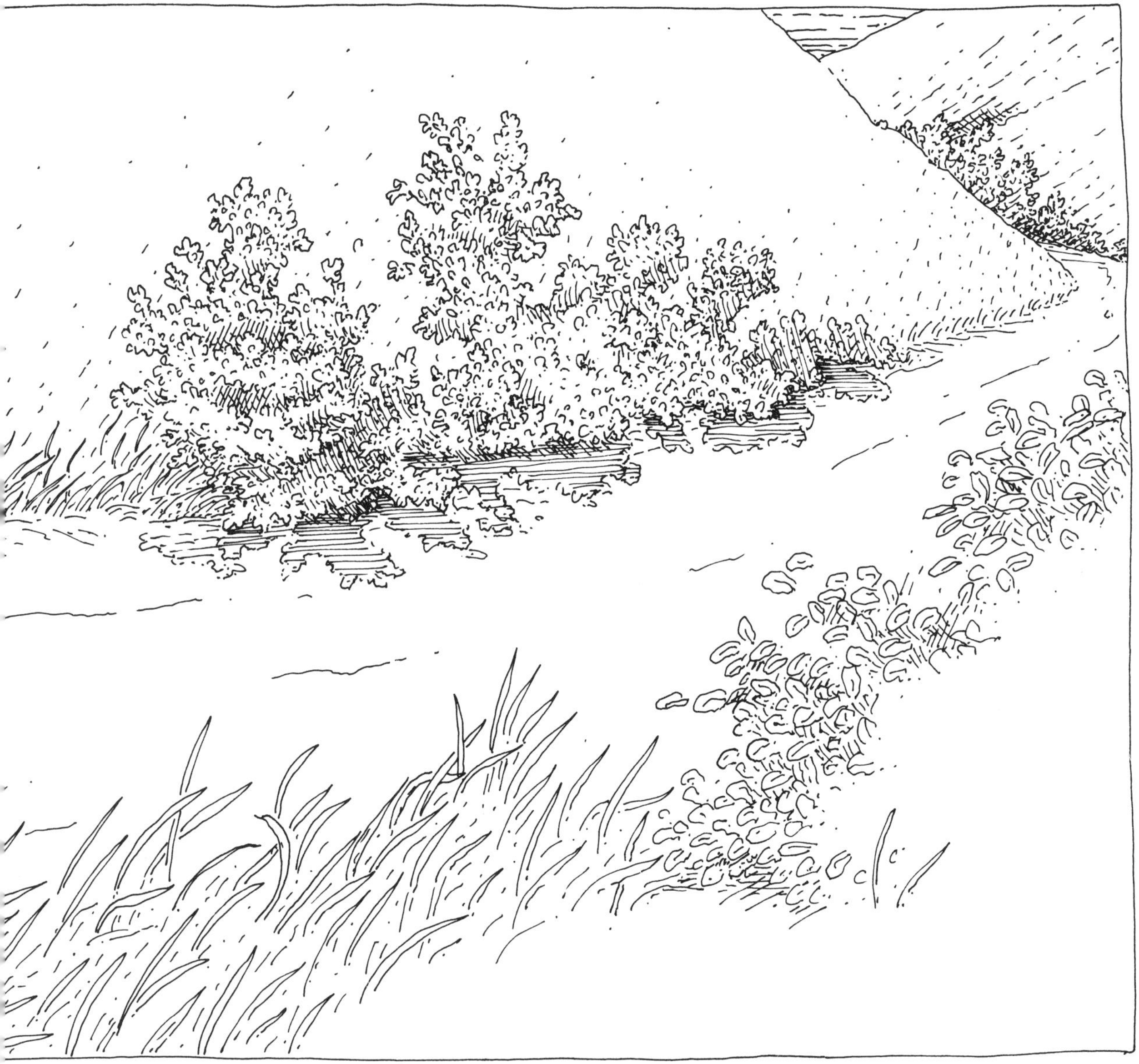

# V
# THE SALT MARSH

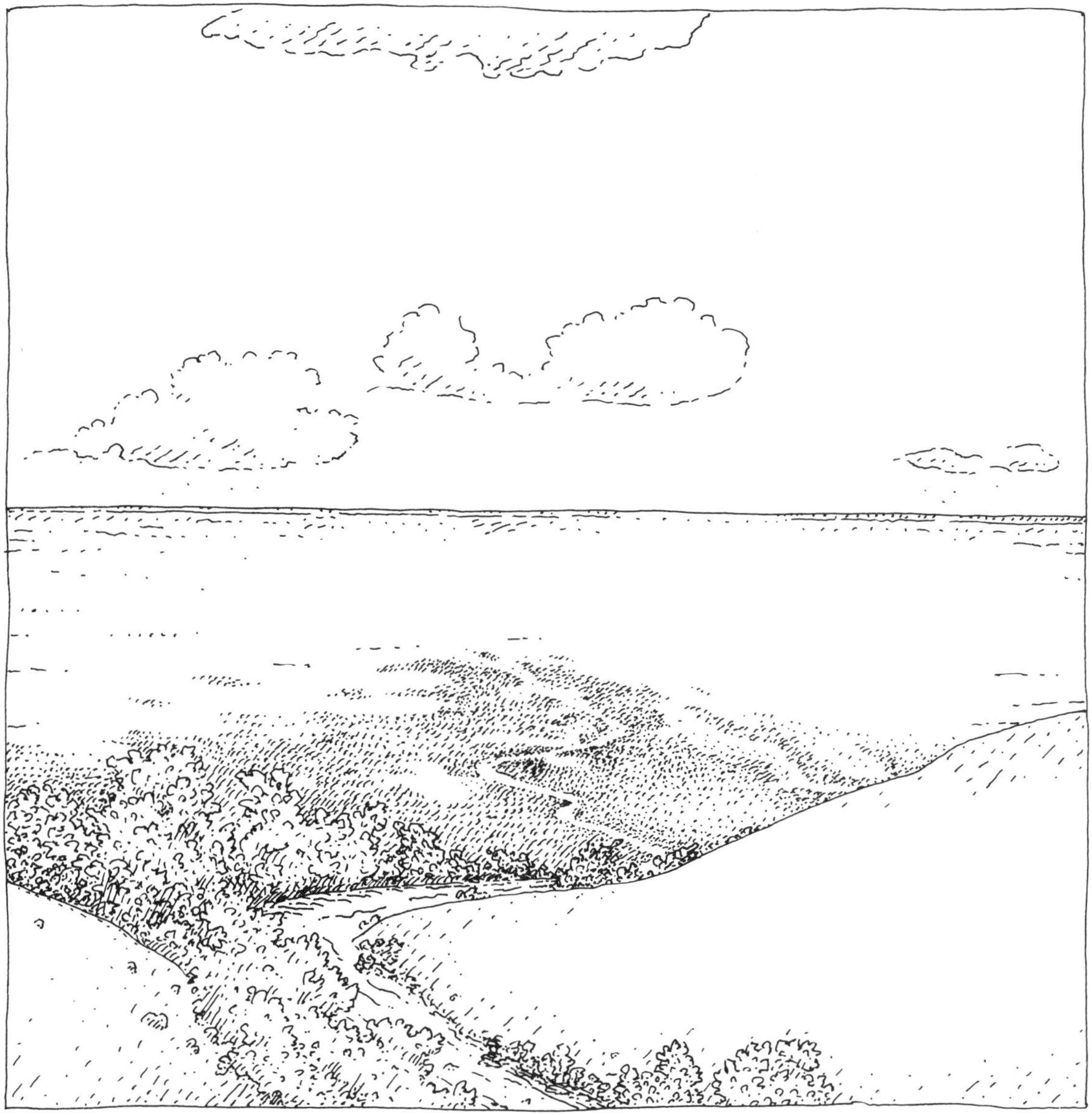

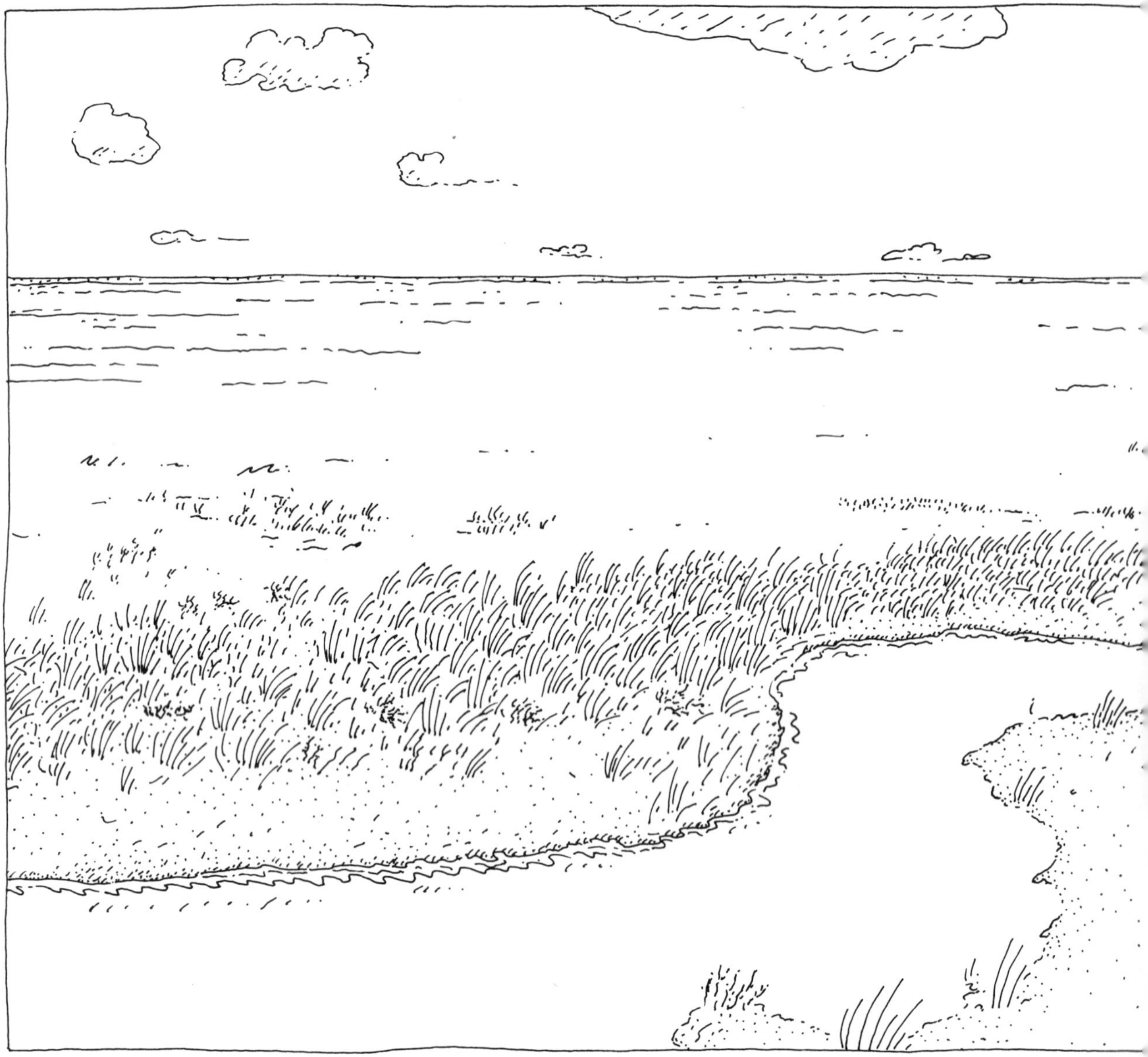

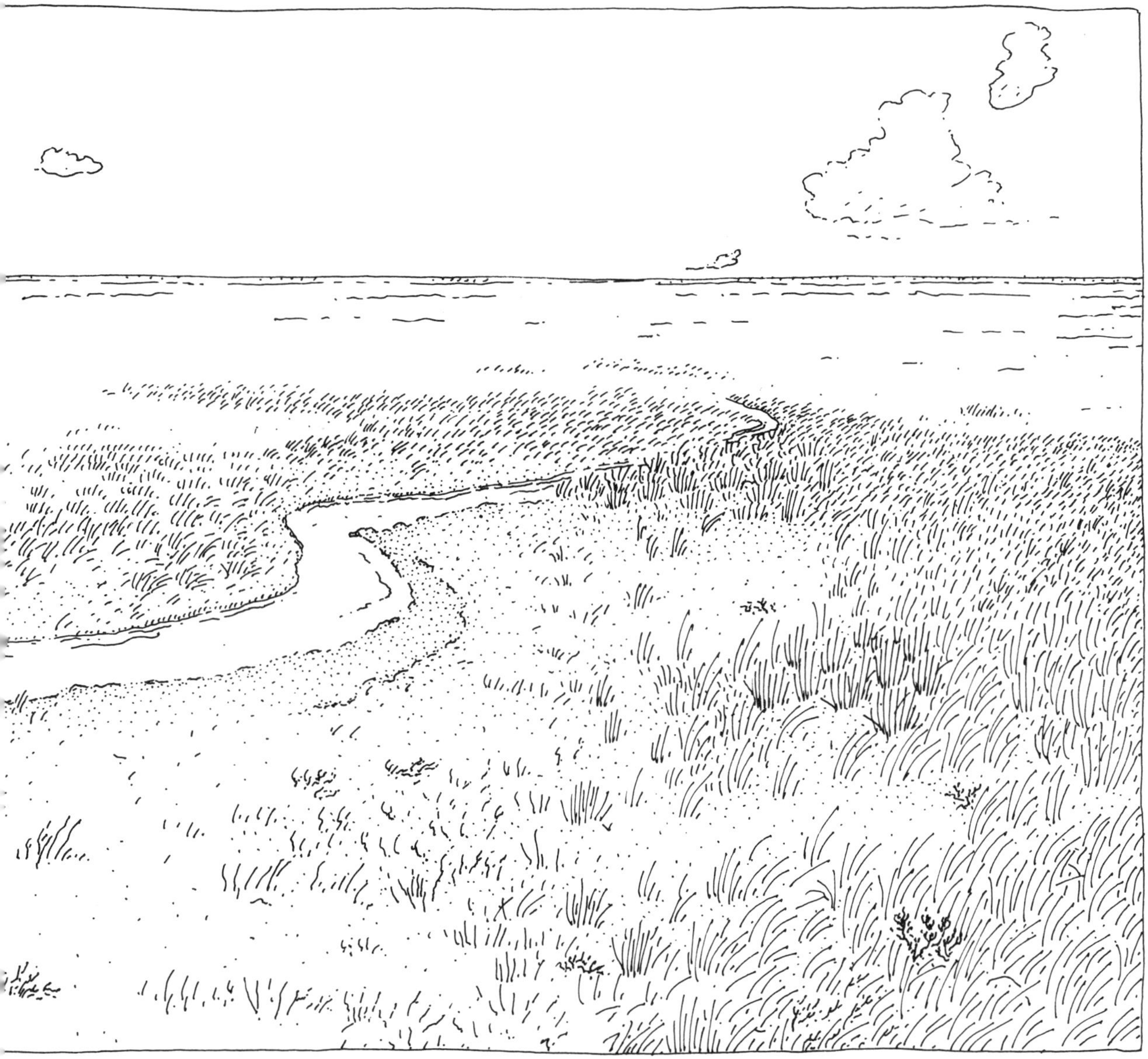

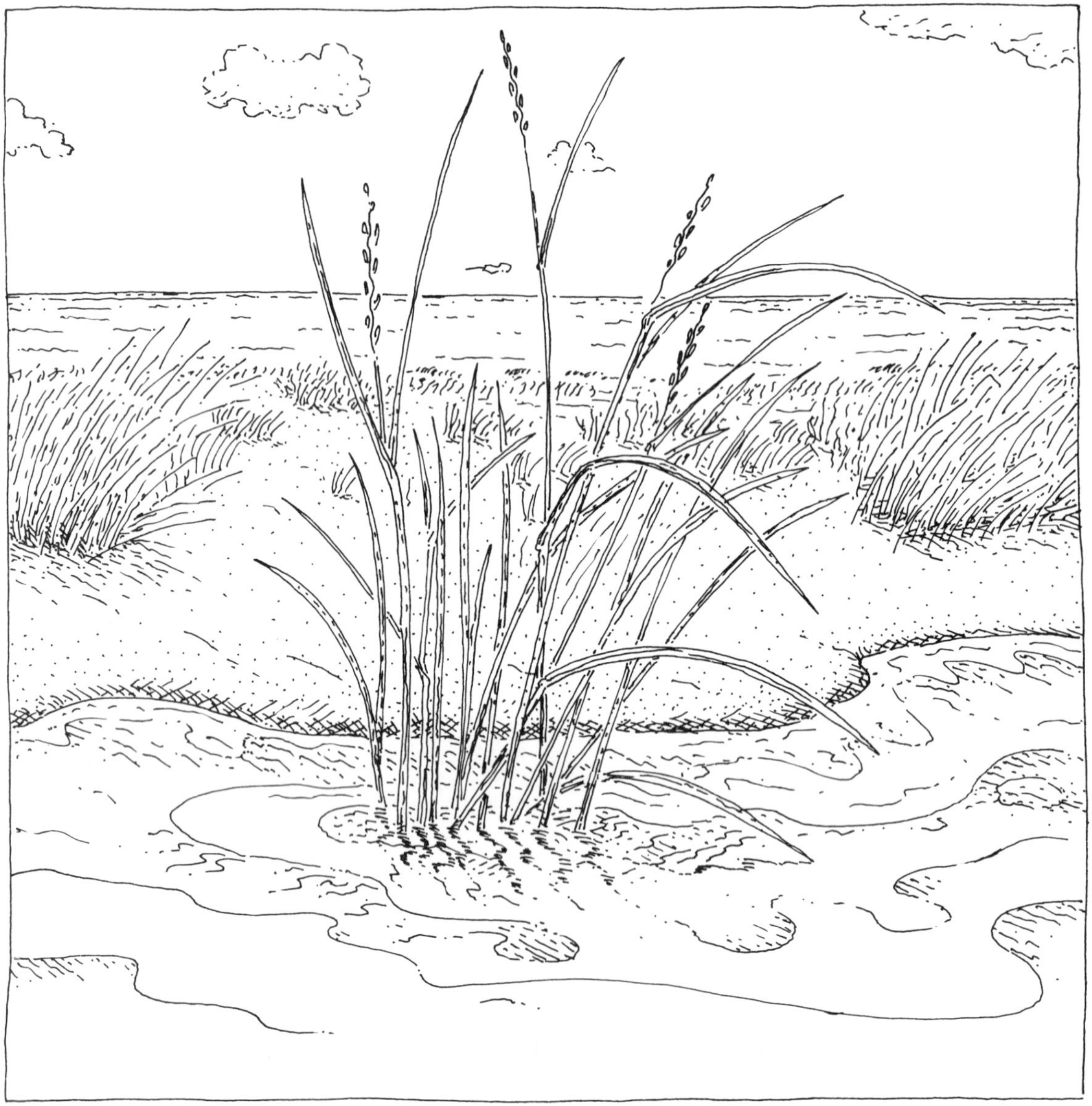

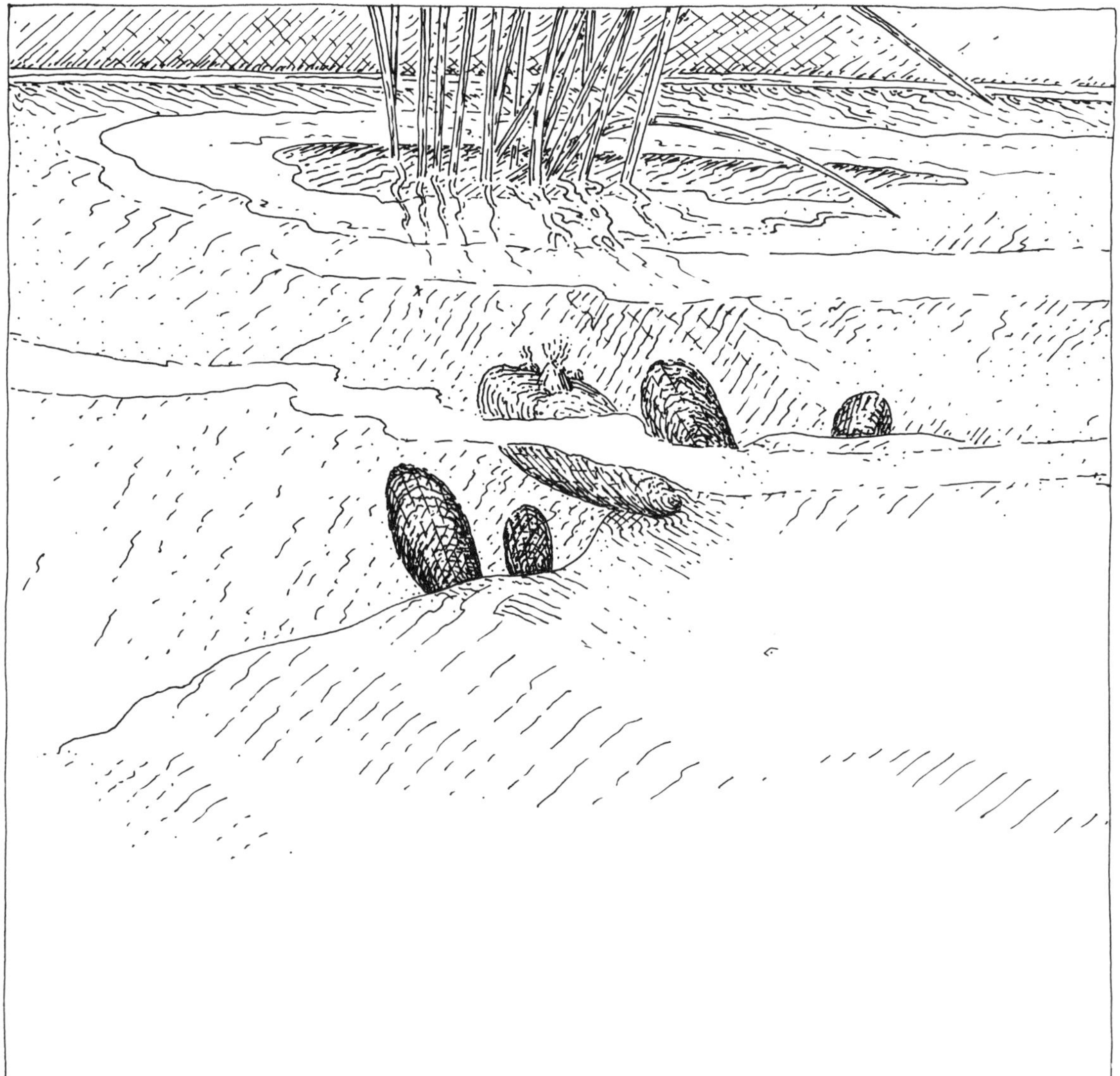

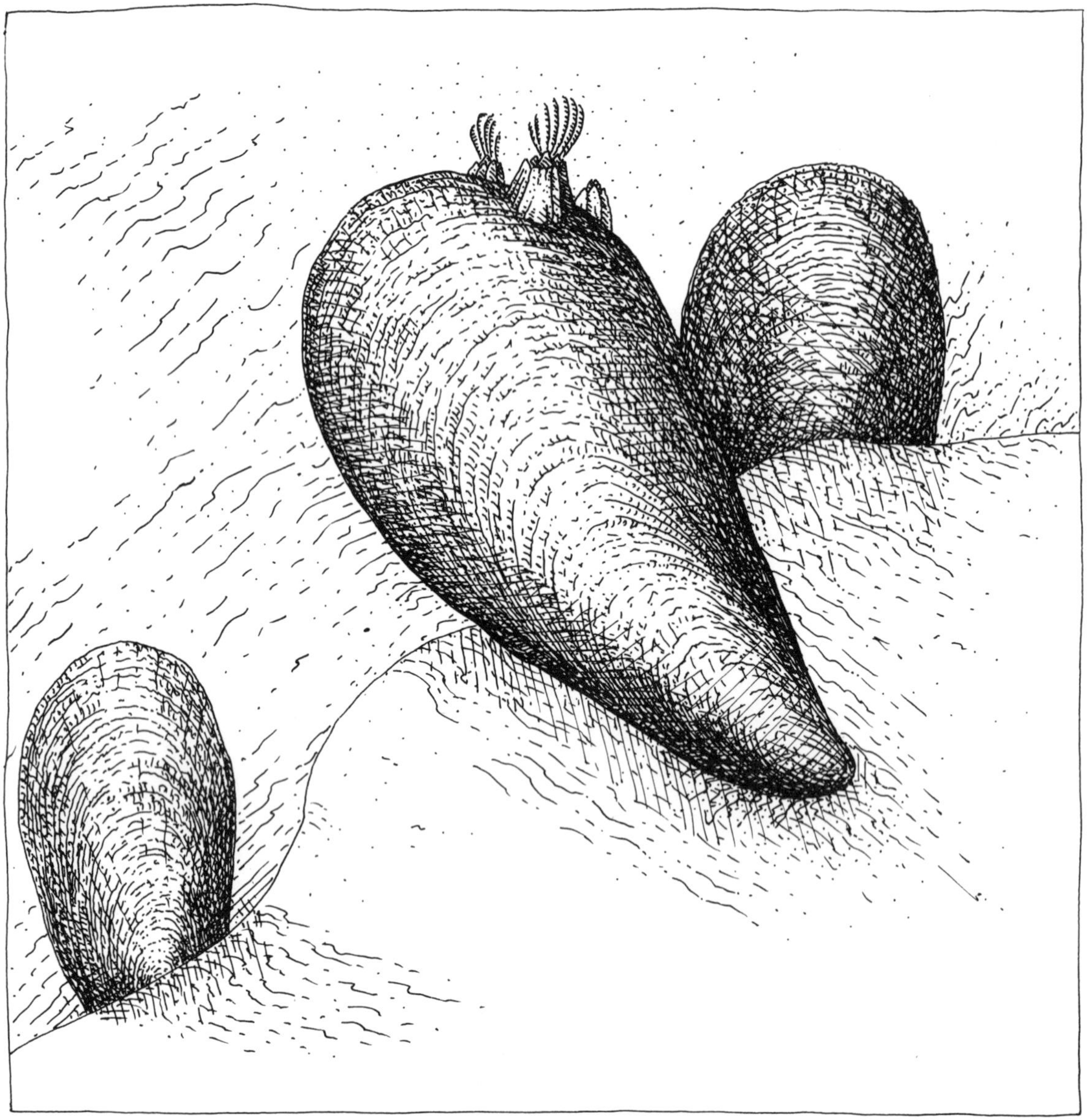

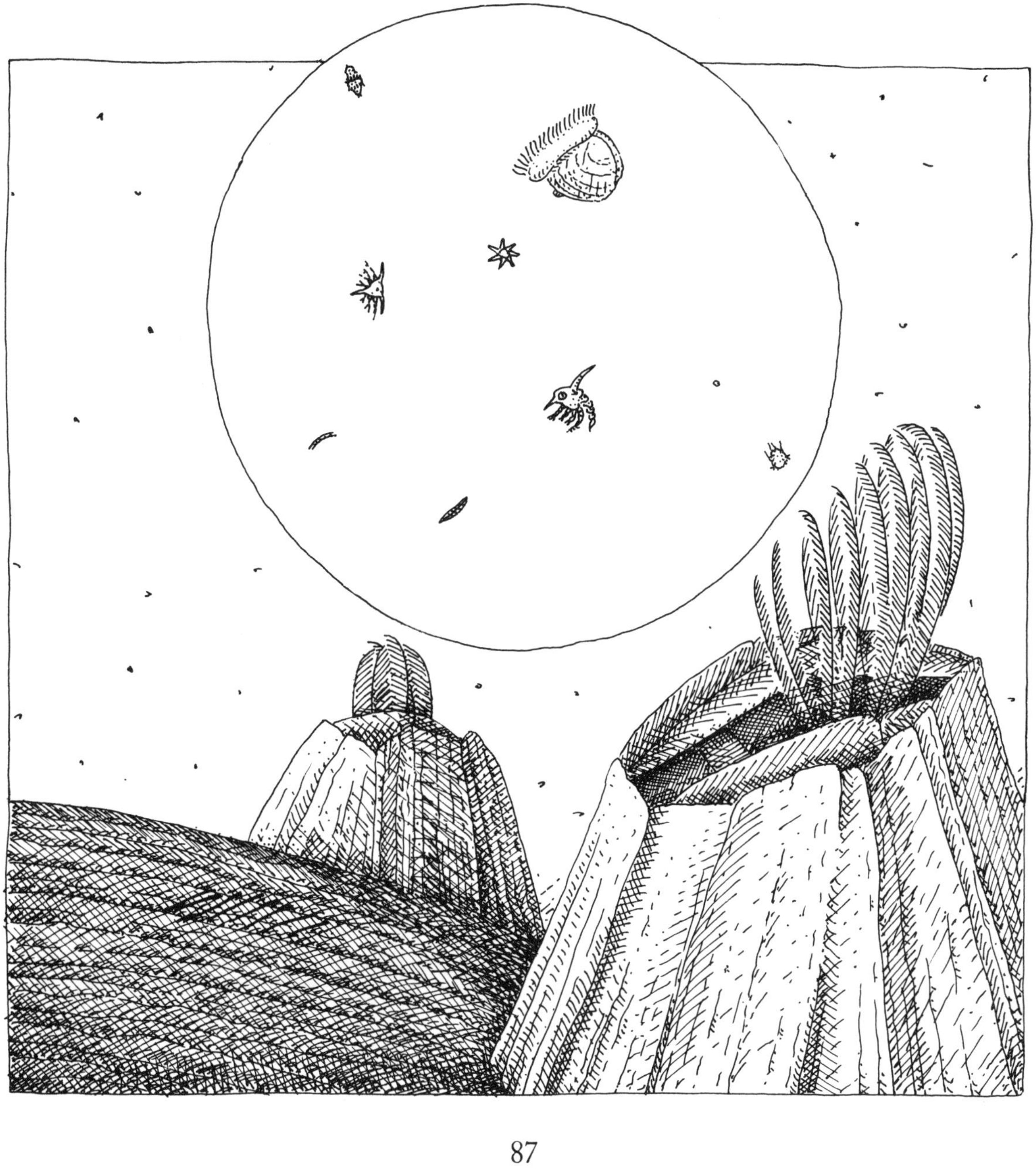

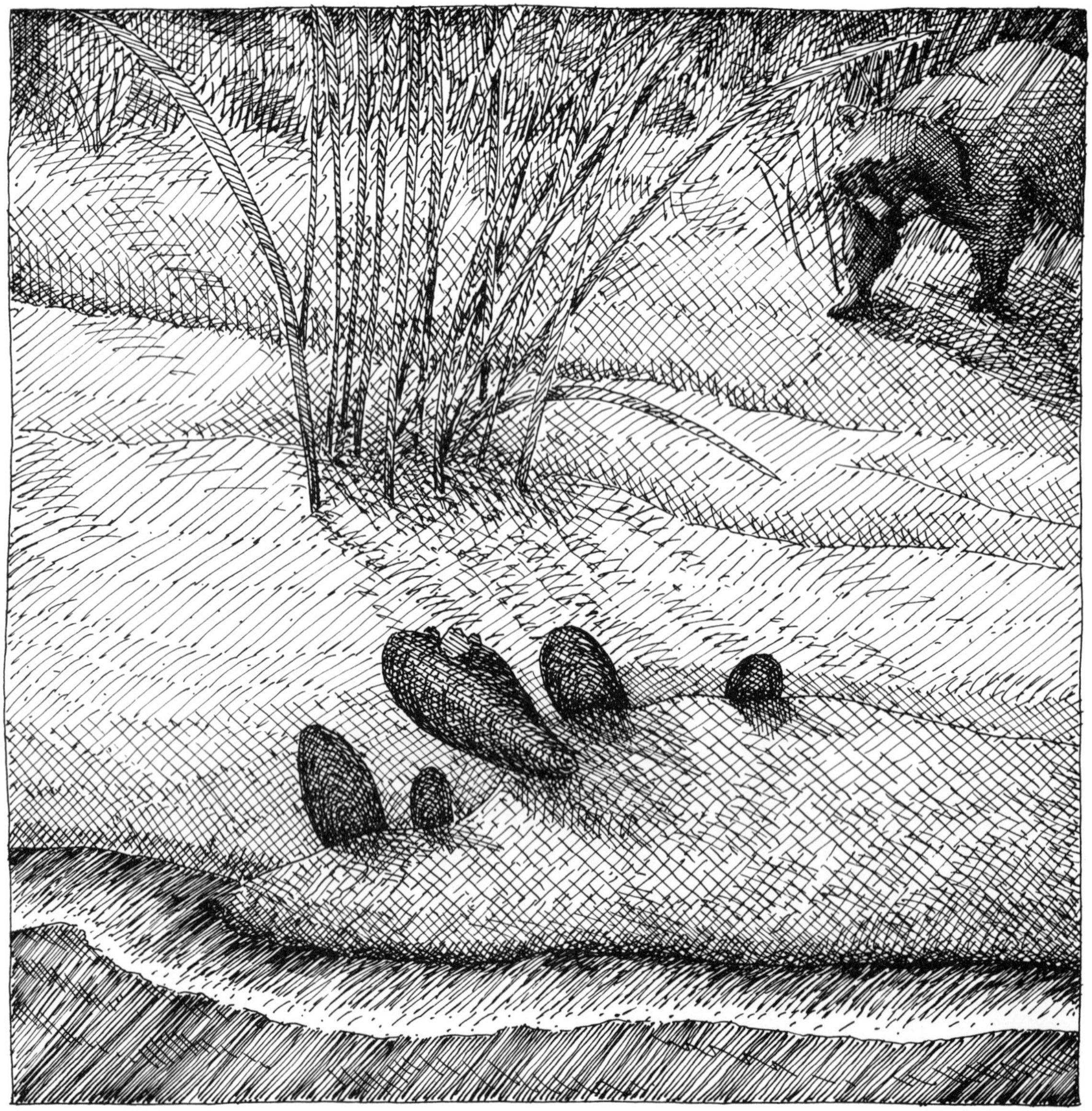

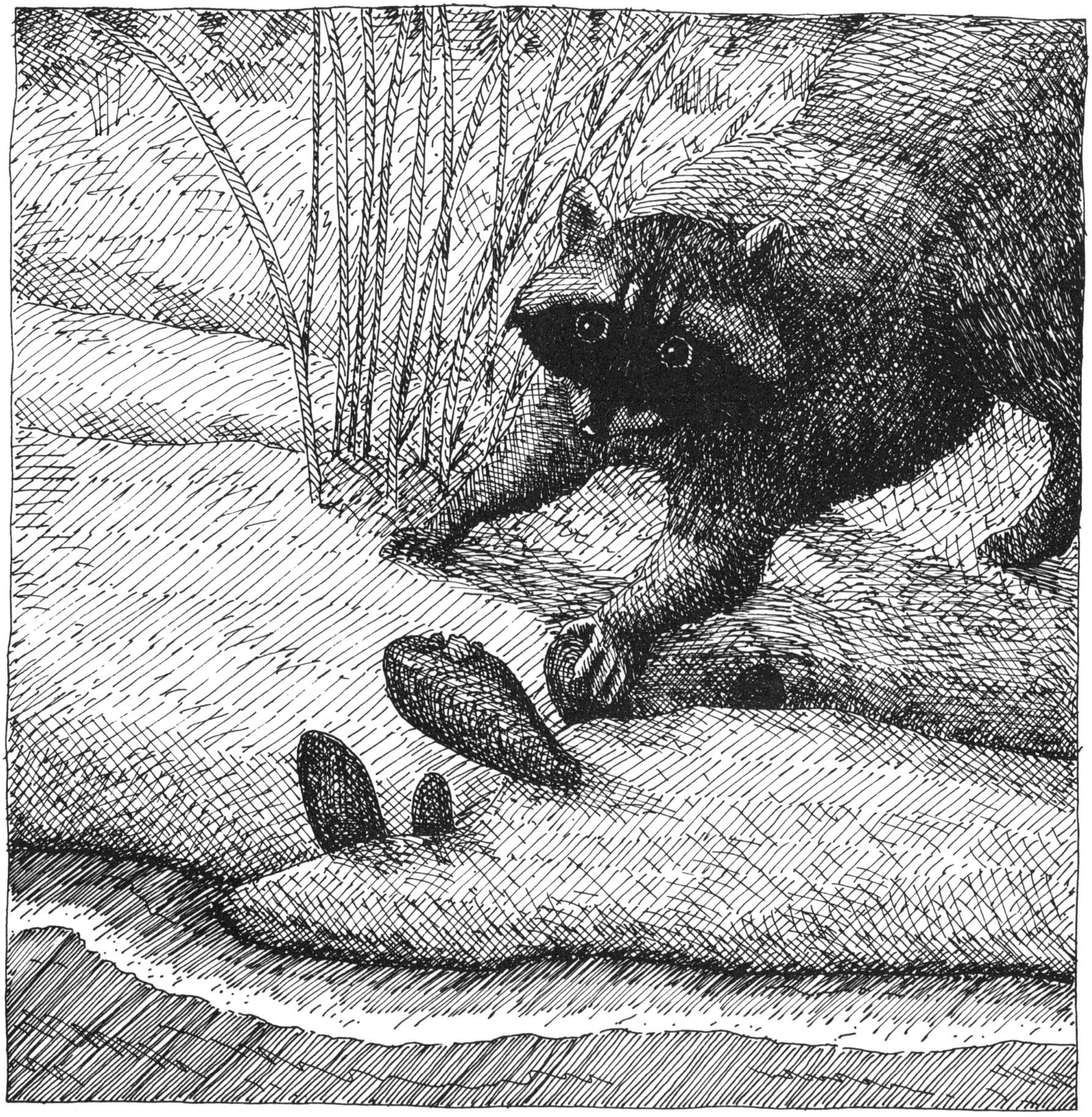

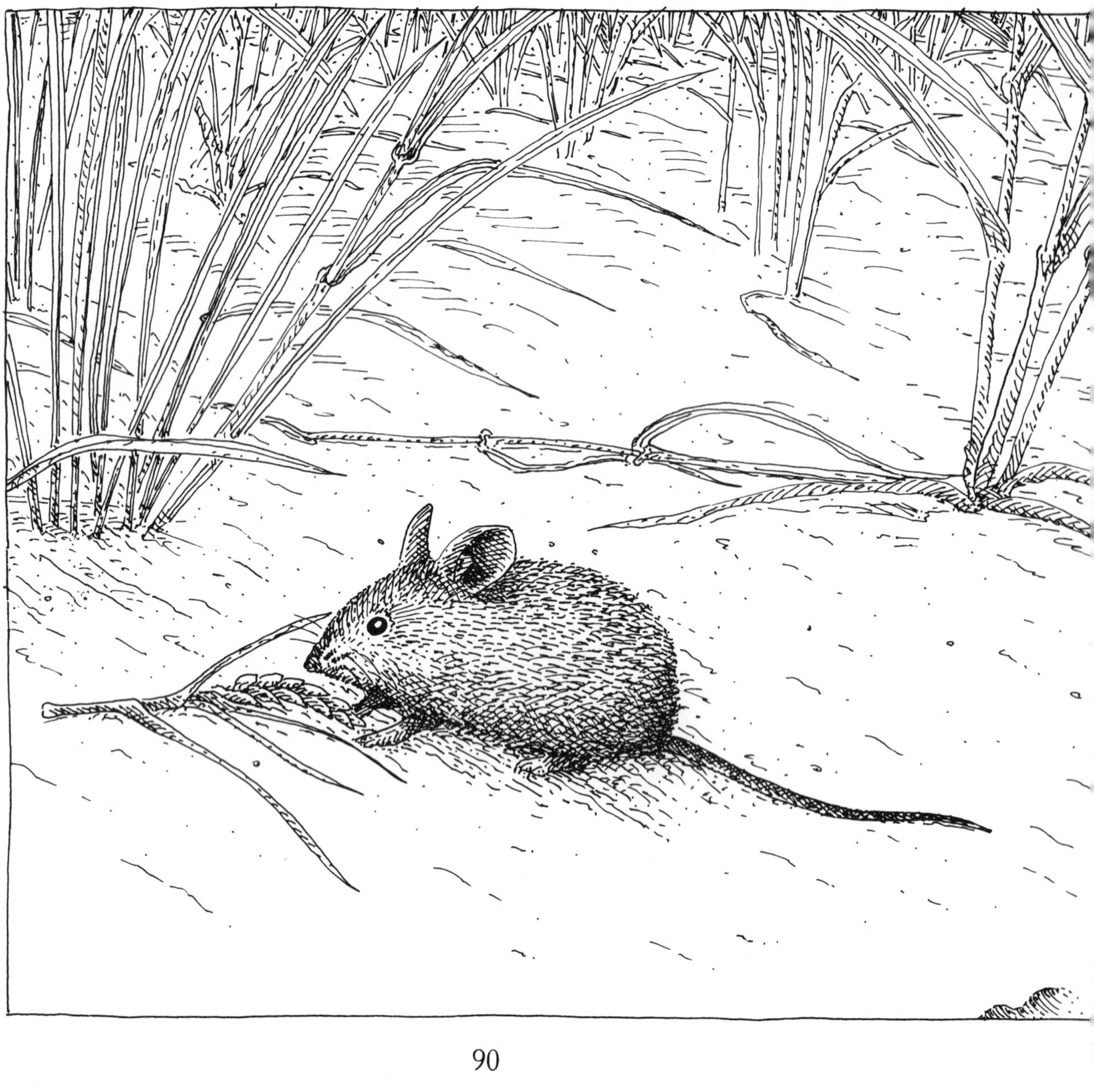

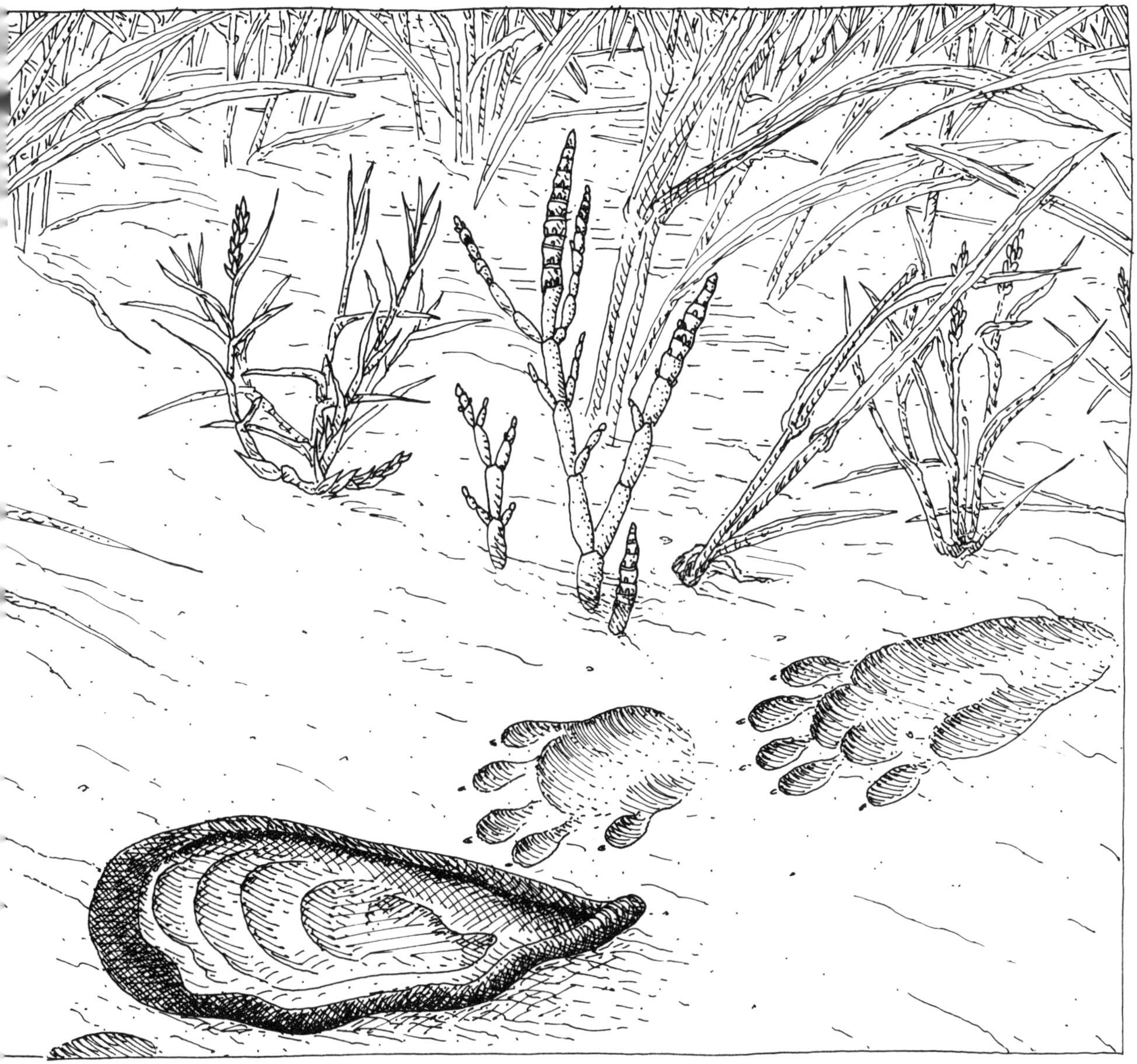

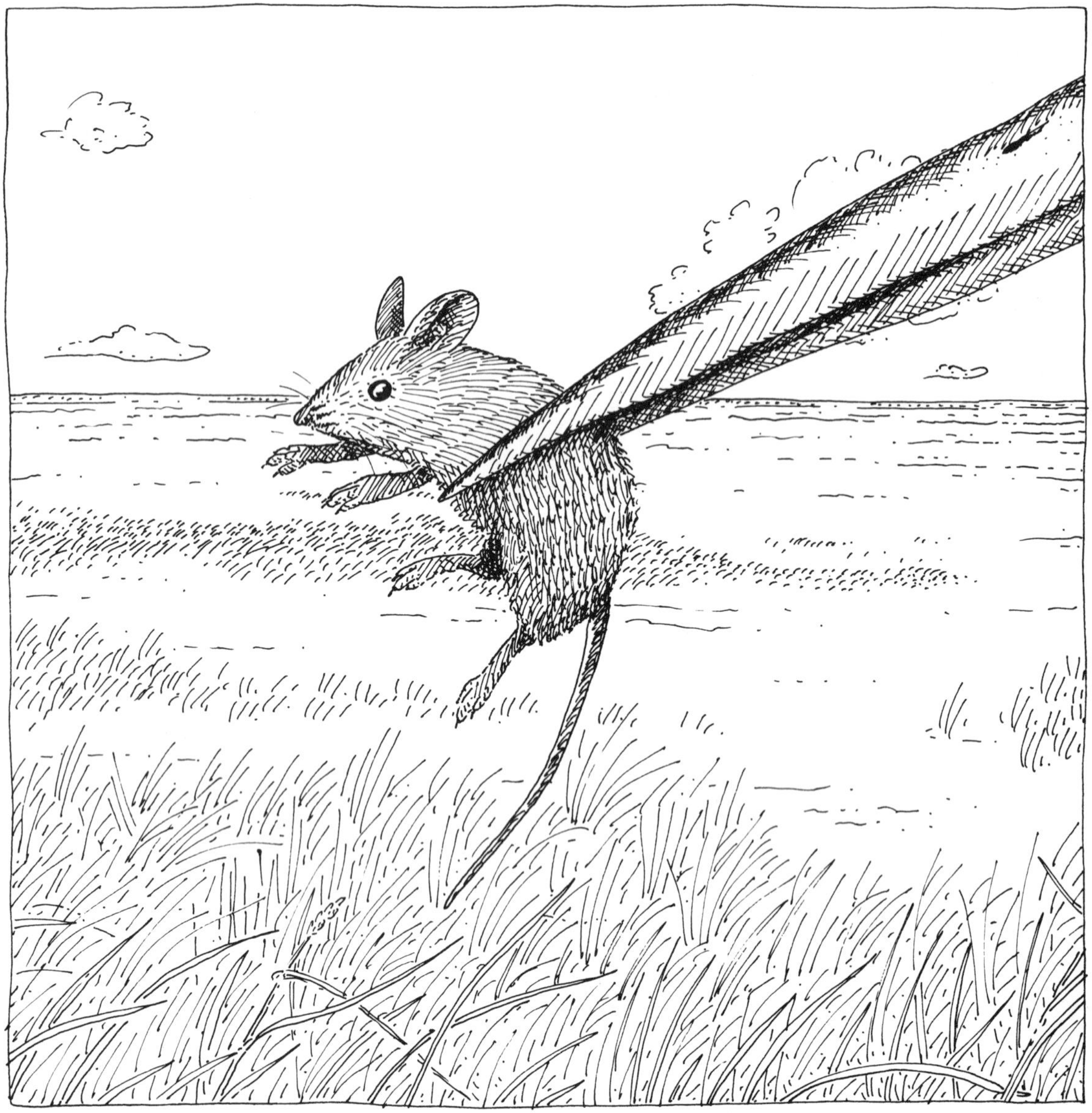

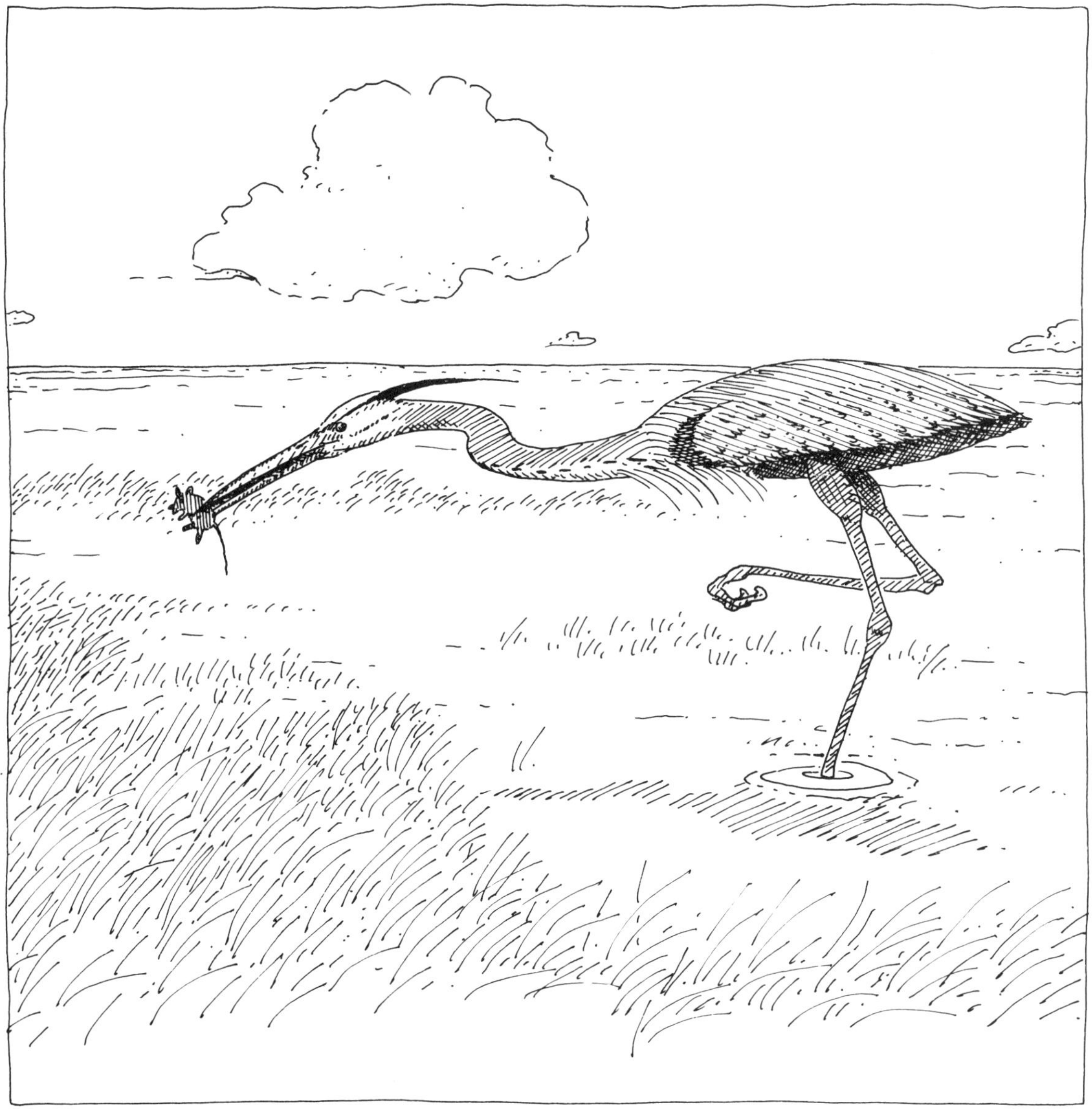

94

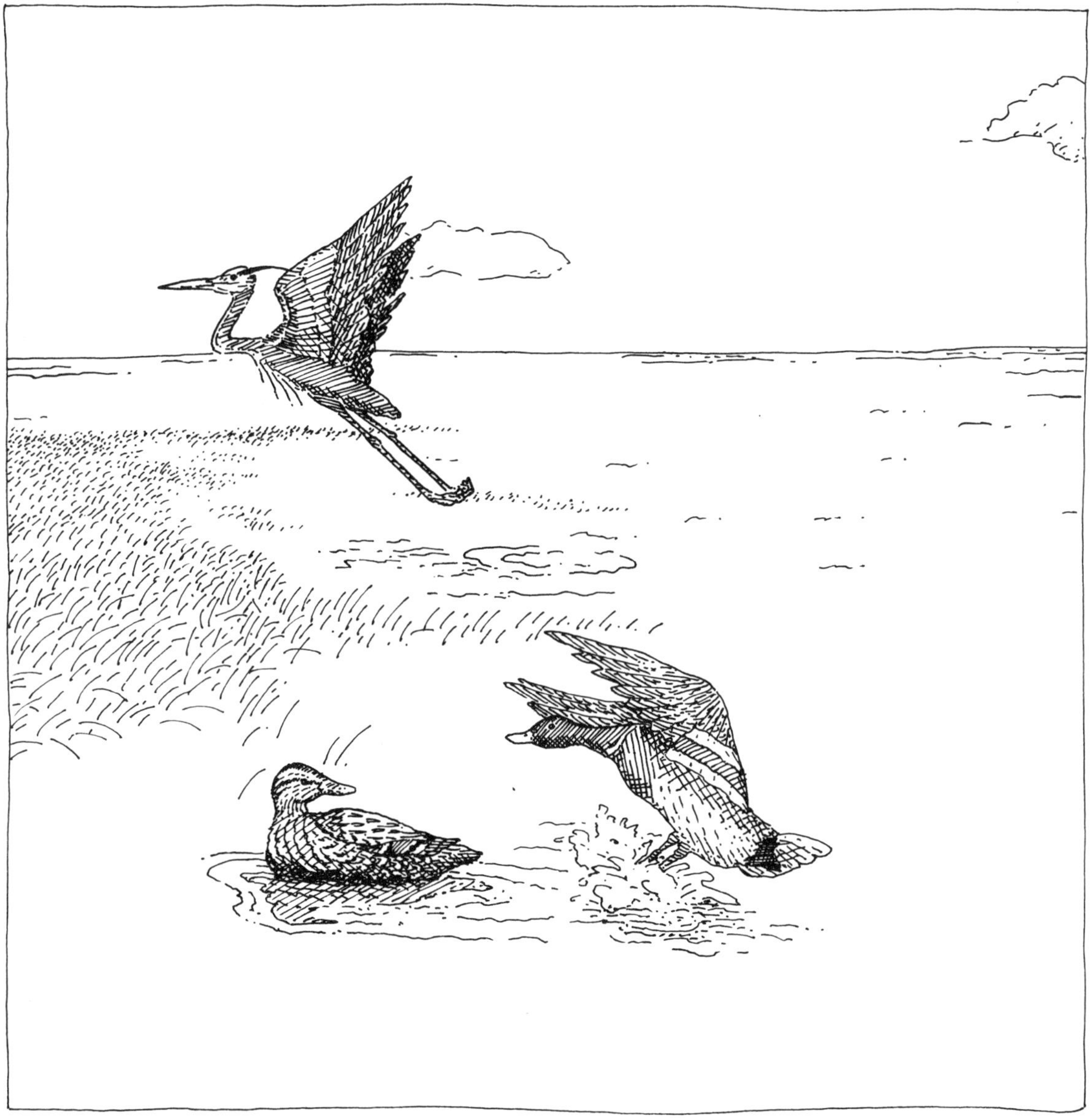

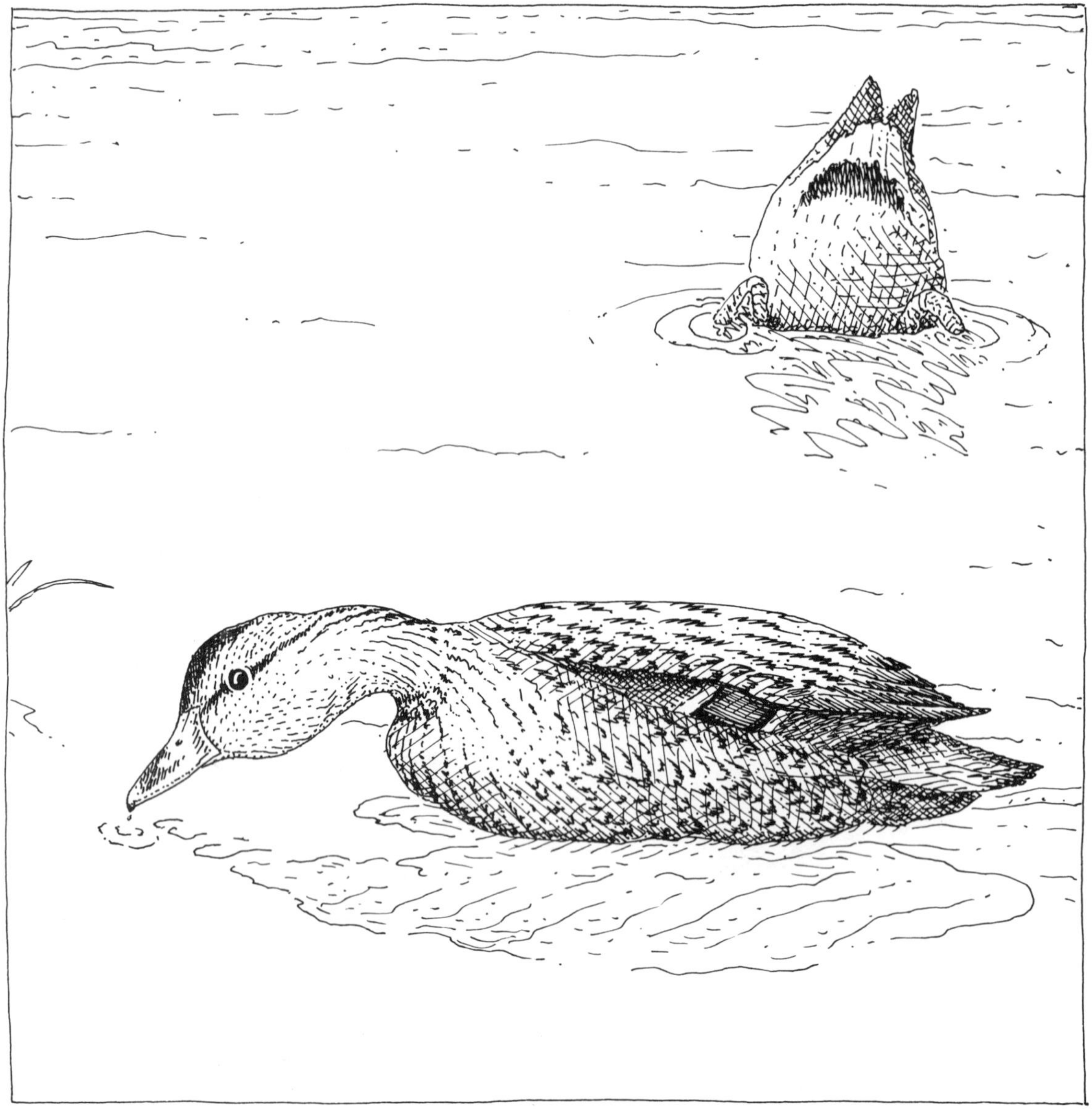

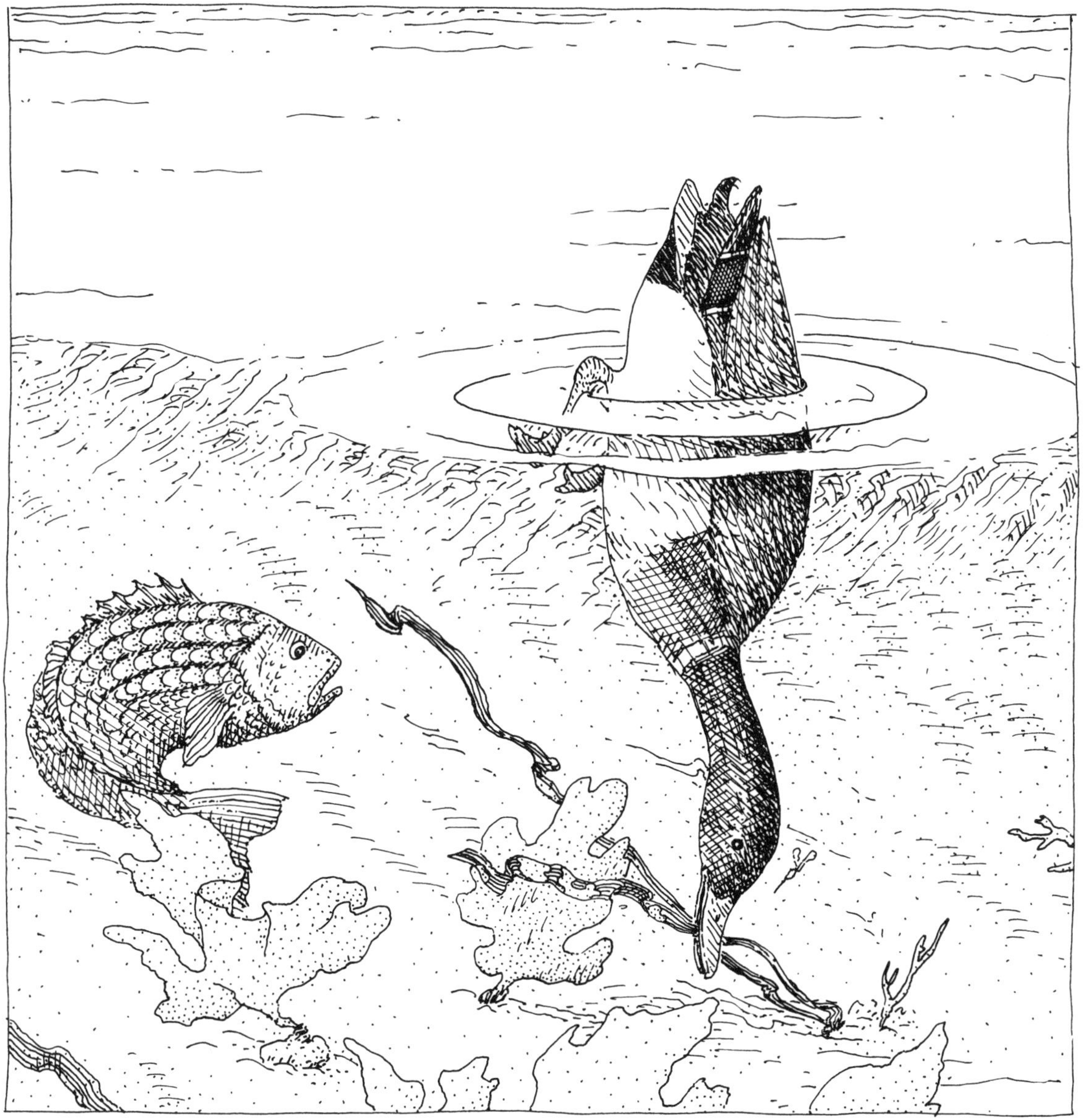

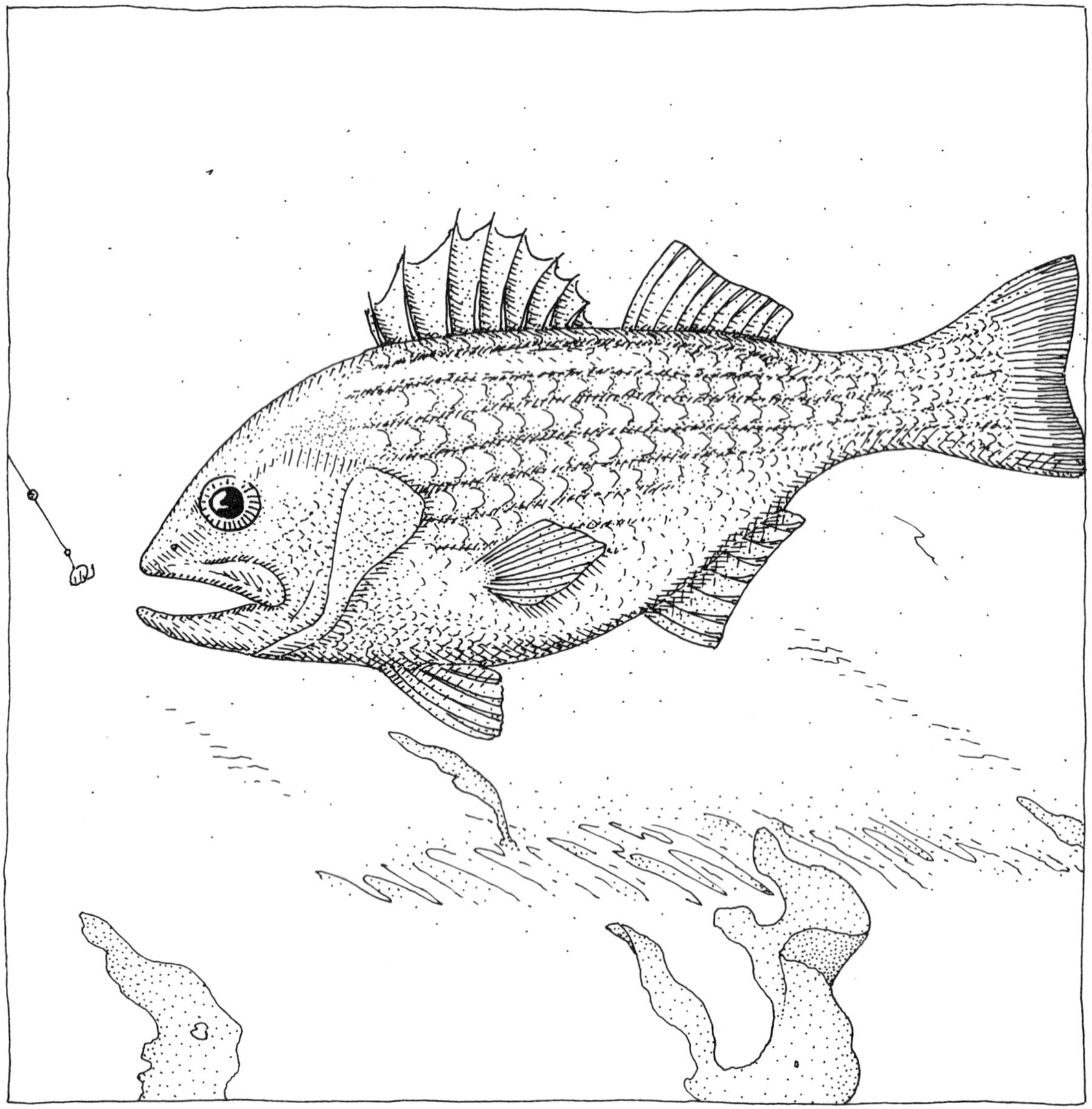

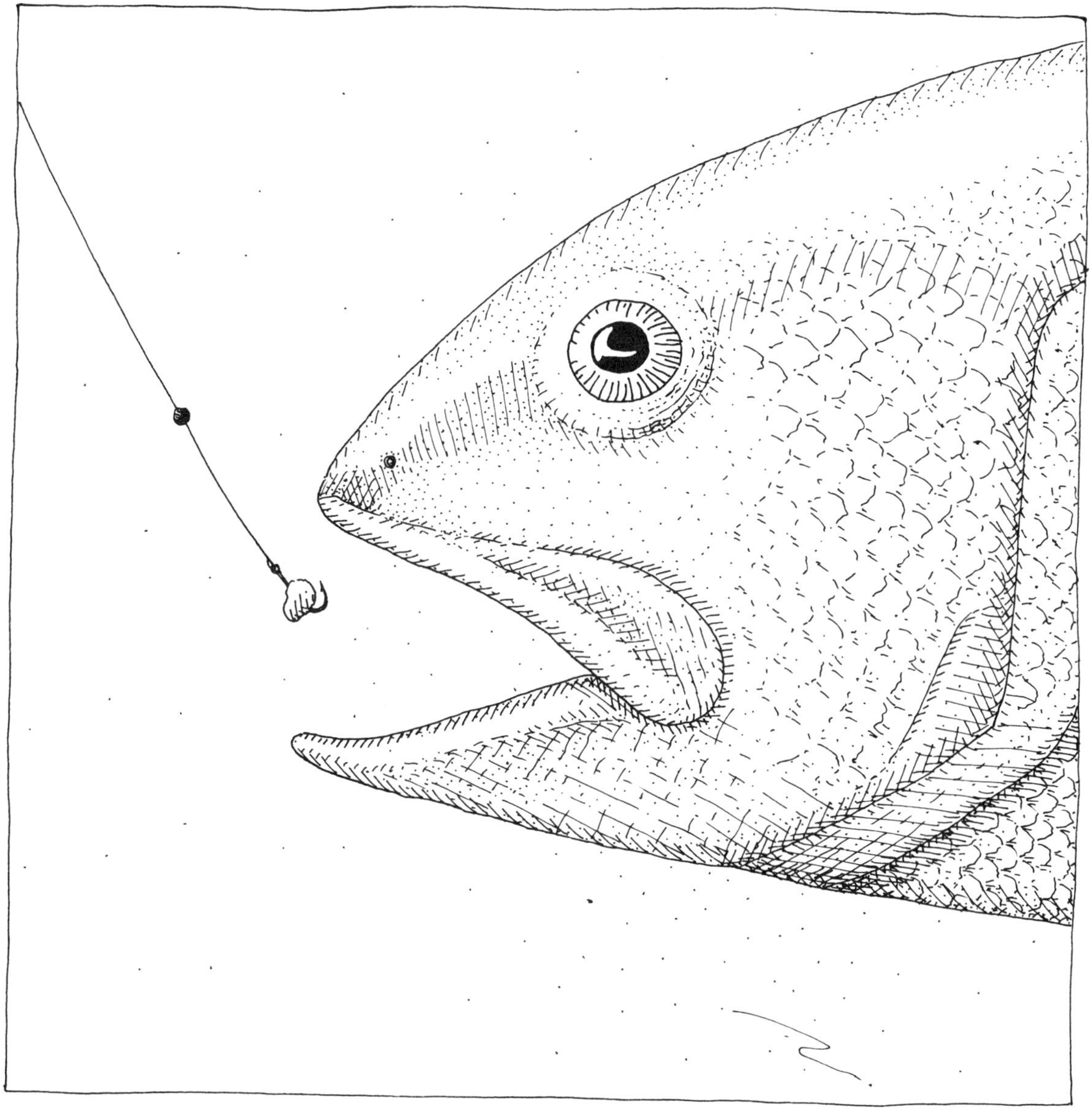

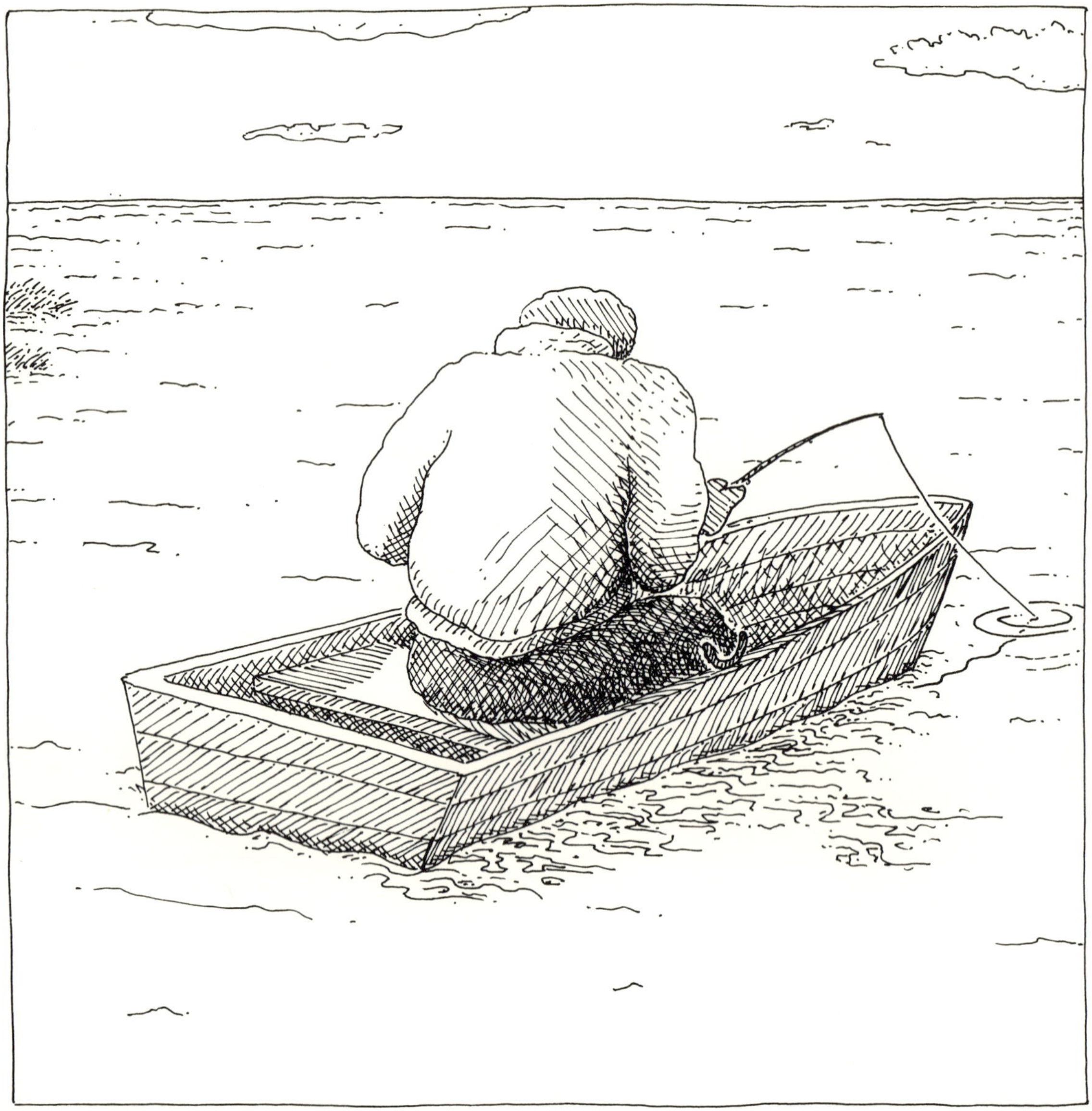

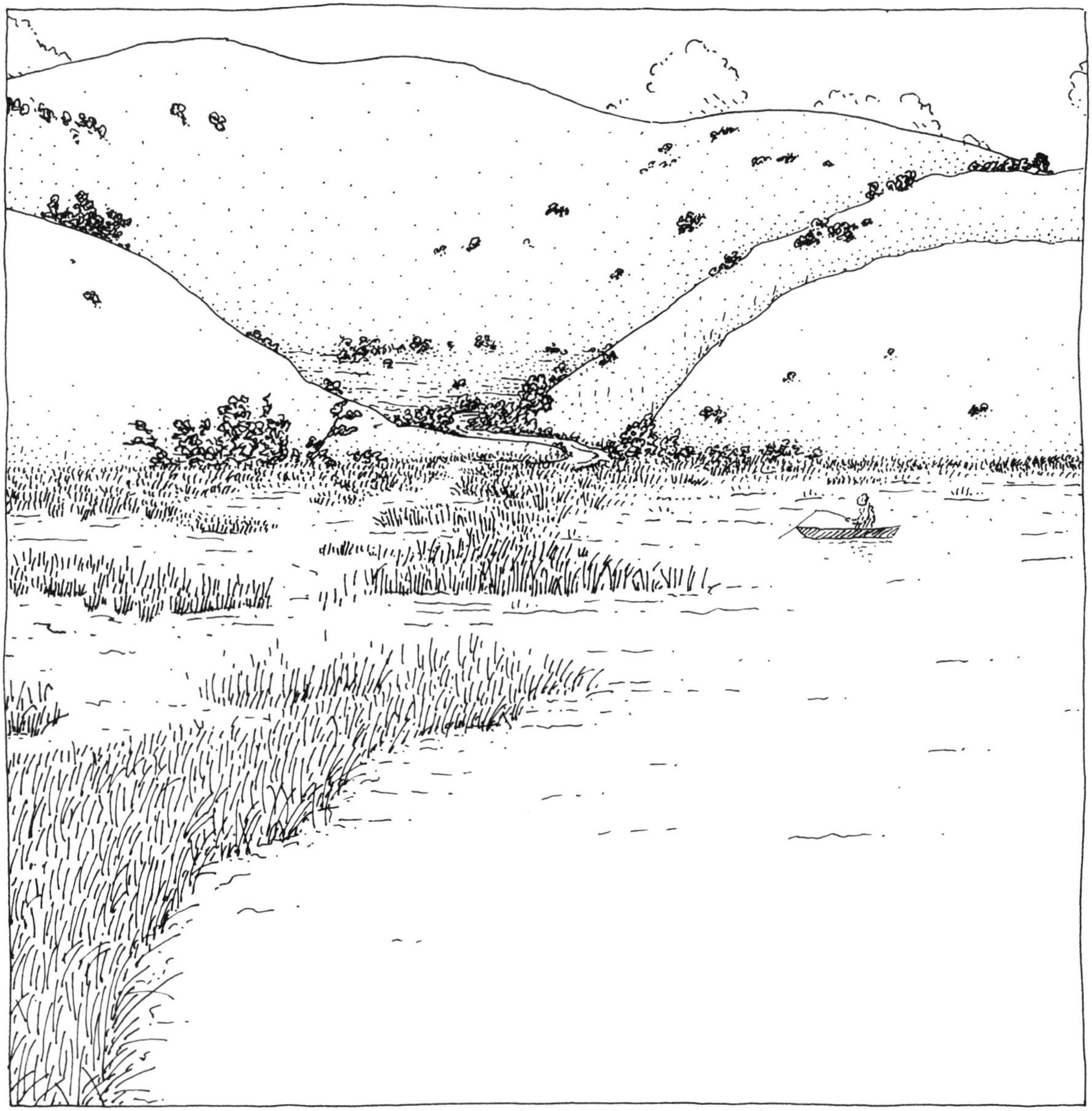

# ABOUT THE STORIES

# I  CRICKET IN THE GRASS

*Grassland Habitat*

  This story takes place in a grassland. Grasses, like all green plants, capture energy from the sun. Light from the sun is what plants use to make food. All animals depend on green plants in one way or another. Animals either feed on plants or they eat other animals that feed on plants. When you visit a grassland habitat, at first you will notice only the grass. But if you look more closely, you will be able to see many other plants and animals living there also.

*The Oak Tree*

    The oak tree provides a home for many creatures that could not survive in the open grass. The bark of the tree is home for insects; the branches provide shelter for birds. The green leaves are food for many creatures, and when the leaves die and fall to earth, they are food and home to still other animals. Trees make the grassland richer by making it possible for more kinds of animals to survive in that habitat.

*The Cricket*

    The cricket's home is deep in the grass. Grass is his shelter as well as his food. Crickets eat the tenderest parts of grasses and other plants. They also eat dead insects. Even though crickets are difficult to see, you have probably heard them chirping. The males rub one wing against the other to make the shrill chirping noise. They do this to attract females.

### The Toad

Just as a cricket must have a forest of grass to live on, a toad must have many insects to stay alive. In addition to crickets, toads also eat bees, flies, ants, beetles, grasshoppers, spiders, and earthworms. Toads are more active at night than during the day. They spend most of their days in holes and burrows. The "warts" on a toad's skin are actually glands that are full of a milky, poisonous fluid. This helps to protect the toad from some of its enemies. However, garter snakes sometimes eat toads in spite of their poisonous warts.

### The Garter Snake

The garter snake is well adapted to survive in the grassland. It is long and slender and can easily slither its way through the plants in search of food. Garter snakes swallow their meals whole. If they are swallowing large animals, they need to open their mouths very wide. Snakes have jaws that unhinge, allowing them to eat animals whole. The meal is digested inside the snake's stomach. Like all reptiles, the garter snake is cold-blooded. Cold-blooded does not mean that its blood is cold, but that its body takes on the temperature of the world around it. That is why snakes lie in the sun to stay warm.

*The Red-Tailed Hawk*

    The hawk hunts by soaring on warm air currents, using as little energy as possible. It sees the smallest movement from great heights. When it spots its prey, it falls to earth using the force of its fall to stun the prey. The hawk grasps its prey with its claws. It kills the prey by snapping the neck with its hooked beak.

# II  THE OAK LOG

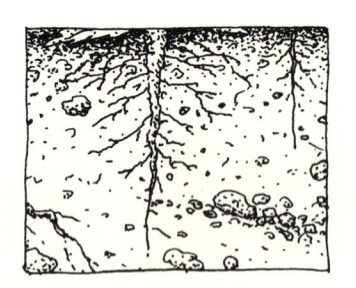

*Soil Habitat*

    This story is about the making of soil. As dead wood and leaf litter decay, they change into soil. Some animals make their homes in the decaying plant matter. They help to break it down into smaller pieces. As plant matter is broken down, it changes as a habitat and provides homes for different animals.

*Leaf Litter*

When plants die or lose their leaves, the plant litter falls to the ground. Animals chew on these leaves, twigs, and branches and grind them down until they are no longer recognizable. Dark, rich soil called humus is made from this decayed plant matter. Humus is a rich environment for growing plants. Although dead plants are broken down year after year, the remains build up very slowly. It takes hundreds of years to build up a thin layer of dark topsoil.

*Dead Wood*

Whole trees are broken down into soil when they die. Burrowing ants, termites, beetles, and many other creatures gnaw tunnels through the wood. These tunnels let air and water deep inside, creating moist, dark homes for special plants called fungi that continue to break down wood.

*Termites*

Termites, like bees and ants, live in huge colonies. At the center of the colony is a queen who lays thousands of eggs. The eggs develop into worker and soldier termites who look different from each other and have different jobs. Workers burrow and dig through the wood. Soldiers protect the colony with their massive jaws. Late in the summer wingless termite workers bore holes in the walls of their home. Winged male and female termites fly from these holes searching for suitable homes for new termite colonies. These females become queens for new colonies. As queens, they do nothing but lay eggs, almost an egg a second during their entire lifetime. Some termite queens live as long as fifty years!

*Fungi*

Fungi are special plants because, unlike green plants, they have no chlorophyll and are not made up of roots, stems, and leaves. Fungi do not use sunlight to make food for themselves like green plants do. Instead they must digest dead and dying plants and animals in order to live. Molds, mushrooms, and toadstools are kinds of fungi that you may have seen growing in damp places after a rain. Most new fungi grow from spores. A spore is like a very simple seed. The seeds of green plants have a hard covering and contain food to help the young plant grow. Spores do not. Fungi release thousands of spores, but only a few of these spores ever find the right conditions for growth. The toadstool or mushroom that we can see is only part of the fungus plant. Below the surface of the soil the mushroom grows a vast web of fibers penetrating wood, leaves, or other food.

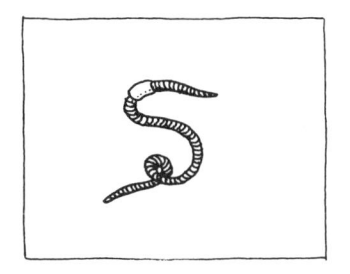

*The Worm*

As wood becomes more like soil, worms burrow into it, eating as they go. A worm has a gizzard which grinds what it eats and sends any food to the stomach. The rest of the matter is deposited at the mouth of the worm's burrow in little piles called castings. Earthworms slowly churn up the soil. They bring deep soil that is rich in minerals up to the surface and bury bits of organic matter deep in the soil where it is broken down by fungi. Earthworms, like fungi, need warmth and moisture in order to live. When the soil is dry or very cold, earthworms burrow deep into the ground and fungi stop growing. During times of dryness and cold, decomposition and soil making slows down almost to a standstill.

*The Mole*

    The mole lives underground. It burrows below the surface of the soil, searching for worms and beetle grubs. It has sharp little teeth and a keen sense of smell, but very poor eyesight. The mole has big front feet with long sharp claws made for digging in the soil. Even its fur, which is short and velvety, is made for the underground life. It allows the mole to move easily backward and forward in its narrow tunnels.

*The Gopher*

    Although the gopher and the mole are both mammals that live underground, they are different in many ways. The gopher eats roots and bulbs rather than worms. It has large front teeth made for gnawing roots. It can close its mouth behind those front teeth so that it can chew on roots without getting dirt in its mouth. The gopher has fur-lined pouches on its cheeks so it can carry food to underground storerooms.

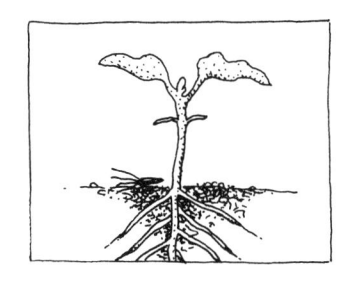

*The Seed*

A seed is very special. Inside it is a tiny plant ready to grow — if it is given what it needs. First, it must fall in the right place. The seed needs moisture and warmth to sprout. Once it sprouts, it needs sunlight, soil, and water. Under the right conditions, the tiny plant inside the seed will begin to swell and crack the outside covering. As leaves grow, so does a small root. The root pushes into the soil, searching for nutrients. In time, a seed will grow into a plant and produce seeds of its own.

# III THE GREEN LEAF CREATURES

*Green Leaf Habitat*

    Many animals make their homes on the leaves, stems, and roots of plants. This story is about that habitat. Plants provide food and shelter for the animals that are adapted to live there. Green plants make their food from the sun's energy and this food flows through the plant as dissolved plant sugars. Insects are often found on the soft, young buds of plants because these areas are easier to eat. Some insects pierce the plant with strawlike mouthparts to suck out the plant's dissolved sugars. Other insects chew up and swallow whole chunks of the plant.

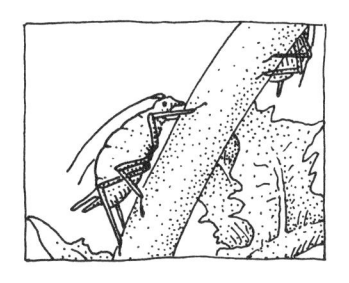

*Aphids*

The aphid is a common insect that makes its home on green plants. Aphids eat the sugary juices of plants by piercing and sucking with their strawlike mouths. Sometimes they make a sweet liquid called honeydew. Ants like to eat honeydew and will even take care of aphids in order to get it. Aphids reproduce during most of the year. Some are born from eggs in the spring. Others are born live in the summer. Even though animals eat aphids, enough aphids survive to lay eggs in the late fall. These eggs will hatch the following spring to start the cycle over again.

*The Ant*

The ant, like the termite, is a social insect which usually builds its nest in the ground. The ant is adapted to survive best living in large colonies. Each colony is the home of one queen and many workers. The queen lays eggs and sometimes lives as long as fifteen years. The workers hunt for food and tend the young. Ants go through a larval and a pupal stage before becoming adults. There are many kinds of ants, and you will find them in many different habitats. Ants like to feed on the sweet honeydew that aphids make. In return for this food, ants do several things to help aphids. They offer aphids protection from predators such as ladybird beetles. During the summer they carry aphids from plant to plant. In winter ants gather aphids and aphid eggs and store them in their nests. The ant and aphid live together cooperatively and benefit each other.

*The Ladybug*

The ladybug, or ladybird beetle, is one of many insects that feed on the slow, fat little aphids. Ladybugs, like all beetles, pass through stages as they grow. They hatch from eggs into larvae that look like little worms. Just like adult ladybugs, the larvae feed on aphids. As the ladybug larva is growing, it has a huge appetite. After some time in the larval stage the ladybug attaches its tail to a leaf and hangs upside down. While it is hanging upside down, it sheds its skin. Underneath is another hard skin that acts as a protective shell. Inside this shell the ladybug is transformed from a larva into an adult. Later the adult ladybird beetle breaks open this shell and crawls out. Unlike most insects, all beetles have hard front wings which cover and protect their bodies. These hard wings are spread out but do not move in flight. Only the soft hind wings are used for flying.

*The Chickadee*

The chickadee is one of many small insect-eating birds that eat aphids. They are often seen hanging upside down on plants looking for aphid eggs or aphids hiding on the underside of leaves. Chickadees will also eat the seeds, nuts, and fruits of many plants. They fly in flocks searching for food. Chickadees nest in holes in decaying stumps or in abandoned woodpecker holes.

# IV  A TURTLE'S WORLD

*Stream Habitat*

The stream is home to plants and animals who have adapted to life in a changing environment. Heavy rains or melting snow can turn a dry stream bed into a raging torrent. Nutrients are washed from the soil with this rush of water. These nutrients provide rich food for algae and tiny (microscopic) water animals. With the coming of summer the stream slows and may eventually dry up. Plants and animals that live in this habitat must be able to adapt to great changes every year.

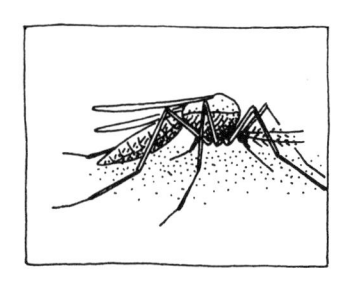

*The Mosquito*

Mosquitos lay their eggs in the calm waters of a stream. Like many insects, mosquitos go through stages before they become adults. A young mosquito, or larva, lives in the water, hanging below the surface. The larva breathes through a tube that pokes above the water. It feeds on bacteria and other bits of food in the water. Before the larva changes to an adult, it passes through a middle stage. Inside a shell-like skin, the larva's muscles dissolve and reform as adult muscles. Wings and other new body parts form. Now the mosquito is ready for a life in the air. Its shell splits, and an adult emerges and flies away. If a mosquito bites you, you'll know it's a female because only she needs blood for her developing eggs. The male feeds on plant juices.

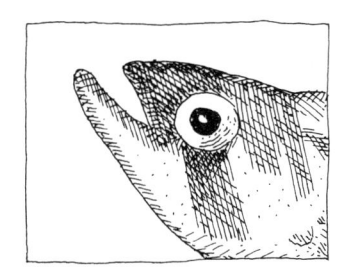

*The Mosquito Fish*

Some minnows feed on algae or small animals at the bottom of streams and ponds. The mosquito fish is called a top minnow. It feeds at the water's surface. This quick little minnow is well adapted for catching the squirming mosquito larvae. One way people have tried to control mosquitos is by putting mosquito fish into streams and ponds where these insects live.

*The Kingfisher*

The kingfisher is one of the few freshwater birds that plunge head first into the water to catch small fish. Its long beak is well adapted for this type of fishing. When the kingfisher makes a successful strike, it flies back to its perch and swallows the fish head first. Like many birds, kingfishers have a defined territory. They fly alone or in pairs along a stream, making a loud rattling cry at intruders.

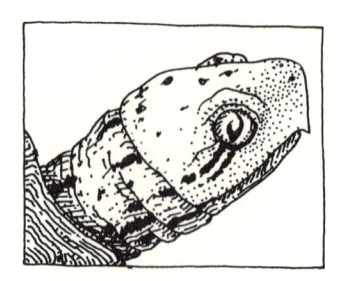

*The Pond Turtle*

Pond turtles are often seen basking on sunny stream banks and on logs. They are seldom far from the water. The pond turtle can withdraw its head, legs, and tail into its shell, but usually it dives and swims to escape enemies. It feeds on frogs, insects, water plants, and dead animals. Turtles are reptiles. Like all reptiles, they lay eggs. The female lays her eggs in sandy areas beside streams and ponds. Baby turtles are on their own from the moment they hatch.

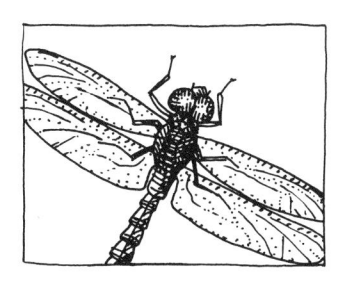

*The Dragonfly*

    This interesting insect is well adapted to survive in the stream environment. The dragonfly has large compound eyes which are like many eyes in one. He can see all around him at one time, making it easy to find his next meal. Because it is a strong, swift flier, the dragonfly usually catches other insects in the air. The dragonfly, like the ladybug and the mosquito, passes through stages of development. First, it is a nymph that lives in the water and breathes through gills like a fish. The dragonfly nymph propels itself by squirting jets of water through its breathing chamber. When the nymph is ready for its change to an adult, it crawls out of the water and sheds its skin. The adult emerges. After its wings have dried it flies away.

# V  THE SALT MARSH

*Salt Marsh Habitat*

    In protected ocean bays and at the mouths of some rivers, salt marshes form. Salt marshes are places where fresh water and salt water mix. Salt marshes begin as mud or sand flats. Algae and flowering plants help to trap more and more silt and sand and the salt marsh grows. Salt marshes usually have one or more streams meandering through them. The stream brings additional silt and sand from higher ground to the marsh. Nutrients wash down over the marsh, then the tides wash them back. This flow of nutrient-rich water is the source of food for the plants and animals of the salt marsh. The salt marsh is a special place because it combines the richness of several habitats: the fresh water, the salt water, and the land. It is the shallow place where these habitats mix to form a new habitat.

*Cordgrass*

Cordgrass is a common salt marsh plant. When it dies and falls into the water, it decomposes into small pieces. This makes food for tiny animals and for the young of larger animals. At some stage of their lives, crabs, mussels, snails, and shrimp feed on this organic matter. These animals lay billions of eggs in the salt marsh. The eggs and young are food for many salt marsh animals.

*Mussels*

Mussels begin their lives as tiny larvae floating in the ocean water in protected places such as salt marshes. Here many of them become food for other animals, but many others find homes on the rocks to begin the next stage of their lives. They attach themselves to rocks and their hard shells grow. Mussels, like clams and barnacles, are filter feeders. As they breathe, water passes over their gills. There are threads on their gills that catch bits of food.

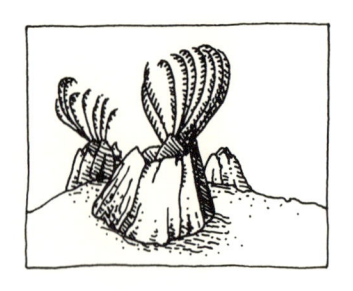

*Barnacles*

    Barnacles also start their lives as free-floating creatures. At some point each barnacle attaches itself by its neck to a solid surface such as a rock or a mussel. Then it grows a cone-shaped shell. As this happens, most of its head dissolves, leaving its featherlike feet to kick food into its mouth.

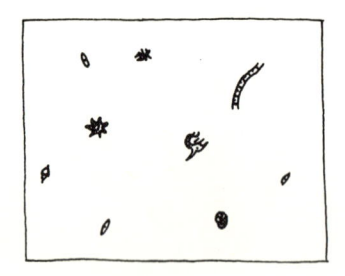

*Plankton*

    The rich salt marsh waters are a fertile home for the young of many ocean creatures. These animals are so tiny that most of them can be seen only with a microscope. They are called plankton. Salt marsh waters also support microscopic plants (one-celled algae) that many of these miniature animals feed on. These plants are also called plankton. All free-floating microscopic life in the ocean, both plant and animal, is called plankton.

*The Raccoon*

    The raccoon is a nighttime visitor to the salt marsh. Mussels, crabs, and clams, even when covered with water, are easy prey for raccoons. The raccoon uses its sensitive hands to search for food by feeling among the rocks, crevices, and salt marsh plants. Outside the marsh the raccoon eats a wide variety of food, including fruits, insects, crayfish, fish, and frogs.

*The Salt Marsh Mouse*

    Deep in the cordgrass lives the salt marsh mouse. Here it makes its tunnels and runways and weaves small grass nests. It relies on the grass to protect it from its many enemies. It is preyed upon by herons and marsh hawks during the day and by owls at night. The salt marsh mouse eats the seeds and greenery of the salt marsh plants.

*The Great Blue Heron*

    This large, graceful bird can be seen hunting in the salt marsh as well as along freshwater streams and marshes. Its long, slender legs hold its body above the water and tall grasses where it can easily see its prey. It catches its prey with a swift jab of its long, sharp beak. Herons eat the fish, frogs, insects, and mice of the wetlands.

*The Duck*

    The duck is often seen feeding underwater with just its tail showing. It is eating small clams, snails, worms, and plants in the underwater mud. Ducks visit salt marshes on their migrations. Like many birds, some ducks fly south to warmer climates in the cold winter months. In the summer they fly north. These migrating birds pick up mud and seeds on their feet. As they move from place to place the seeds drop off on new shores, encouraging the spread of salt marsh plants.

*The Bass*

Some bass live in the ocean, but the female lays her eggs in freshwater streams in areas called spawning grounds. On her way from the ocean to the stream, she passes through the salt marsh. After the eggs hatch, when the young are returning to the ocean, they linger in the rich salt marsh waters, where they grow strong enough to live in the open sea.

*The Fishing Person*

There are some animals who are adapted to survive in many habitats. In the salt marsh, one of the animals sits in a small boat catching fish with a line and bait. But this animal is just as apt to be found in a wooden house near a grassland or in an apartment in a large city. Human beings do not often think of themselves as animals.

They are not always aware of how they affect the habitats of other animals or of plants. But, like all the other creatures in this book, humans have living places and depend on other creatures and plants for their food and other things they need.

What is your living place? What plants and animals live near you? Which ones are used to feed you or to make your clothes? How many habitats are changed by your presence? Try to look beyond your own home for the answers to these questions.